Undergraduate Lecture Notes in Physics

Undergraduate [illegible] and applied physics. Each title in the series is [illegible] instruction, typically containing practice [illegible] summaries, and suggestions for further [illegible]

[illegible] of the following:

- An exceptionally clear and concise treatment of a standard undergraduate subject.
- A solid undergraduate-level introduction to a graduate, advanced, or non-standard subject.
- A novel perspective or an unusual approach to teaching a subject.

[illegible] especially encourages new, original, and idiosyncratic approaches to physics teaching at the undergraduate level.

The purpose of ULNP is to provide intriguing, absorbing books that will continue to be the reader's preferred reference throughout their academic career.

For further volumes:
http://www.springer.com/series/8917

Undergraduate Lecture Notes in Physics (ULNP) publishes authoritative texts covering topics throughout pure and applied physics. Each title in the series is suitable as a basis for undergraduate instruction, typically containing practice problems, worked examples, chapter summaries, and suggestions for further reading.

ULNP titles must provide at least one of the following:

- An exceptionally clear and concise treatment of a standard undergraduate subject.
- A solid undergraduate-level introduction to a graduate, advanced, or non-standard subject.
- A novel perspective or an unusual approach to teaching a subject.

ULNP especially encourages new, original, and idiosyncratic approaches to physics teaching at the undergraduate level.

The purpose of ULNP is to provide intriguing, absorbing books that will continue to be the reader's preferred reference throughout their academic career.

Ganesan Srinivasan

Life and Death of the Stars

Ganesan Srinivasan
Bangalore
India

Originally published by Universities Press (India) Private Limited 2011, ISBN 9788173717420

ISSN 2192-4791 ISSN 2192-4805 (electronic)
ISBN 978-3-642-45383-0 ISBN 978-3-642-45384-7 (eBook)
DOI 10.1007/978-3-642-45384-7
Springer Heidelberg New York Dordrecht London

Library of Congress Control Number: 2013956803

Printed on acid-free paper

Springer is part of Springer Science+Business Media (www.springer.com)

Foreword to the Edition Published by Universities Press

Lord Martin Rees
Professor of Cosmology and Astrophysics
Astronomer Royal
Master of Trinity College, Cambridge
Past President, Royal Society

If you chose 10,000 people at random, 9,999 would have something in common—their business and their interests would lie on or near the Earth's surface. The other would be an astronomer. I'm lucky to be one of this strange breeds—as is Dr. G. Srinivasan, the author of this series of monographs entitled *The Present Revolution in Astronomy*. But astronomy isn't just for astronomers. Its findings are fascinating and it is as important to understand the cosmos as it is to appreciate the rest of nature. The entire cosmos is part of our environment. Indeed, the dark night sky is one feature that's been essentially unchanged throughout all human history, shared by all cultures—though it has been interpreted in many different ways.

Astronomers are the heirs to a long tradition. Astronomy is the oldest science—except perhaps for medicine. Its origins lie in the need to establish a calendar, to measure time, and to interpret the patterns and regularities seen in the sky. Our knowledge is now advancing faster than ever before—thanks to powerful telescopes and probes that travel to other planets. A wide public has shared the excitement of this vicarious exploration.

We can't send actual probes beyond our solar system, but with our telescopes, we can study stars in detail. In the last decade, we have learnt something that's made the night sky far more interesting. Stars aren't mere twinkling 'points of light'. They're orbited by retinues of planets, just like the Sun is. Some of these planets may be like the Earth—but whether there is life on any of them is a question that challenges future generations of scientists.

We have come to realise the immense scale of the universe—in both space and time. We live in a galaxy containing more than a hundred billion stars; but this galaxy is itself just one of a hundred billion visible with modern telescopes. By looking far away in space, we can penetrate far back in time, because the light from distant objects took a long time to reach us. Astronomers have an advantage over geologists and fossil hunters: they can actually observe the past and trace cosmic history right back to the formation of the first stars and galaxies. Indeed, there is compelling evidence that our universe is the expanding aftermath of a "big bang" nearly 14 billion years ago.

We have learnt one crucial thing about the universe: it is governed by physical laws that we can understand, and these laws seem to be the same everywhere. By analysing the light from a distant galaxy, we can infer that the atoms it's made of behave just like those we study in the laboratory. It's because of this uniformity that we can understand the structure of stars and their life cycles, and how, from simple beginnings, stars, galaxies and planets emerged to form the complex structured cosmos of which we are a part.

The cosmos is a unity. There are links between the very small—the microworld of atoms—and the very large—stars and galaxies. Stars form, evolve and die (sometimes explosively). They are powered by nuclear fusion—a controlled version of what happens in a hydrogen bomb. Over their lifetime, this process generates, from pristine hydrogen, atoms of carbon, oxygen and iron. All the atoms on Earth, and in our bodies are the ashes from long-dead stars. We are the 'nuclear waste' from the fusion power that makes stars shine. To fully understand ourselves and our origins, we must understand not only Darwinian evolution, but also the atoms all life is made of, and the stars that made those atoms. This wonderful story should be part of everyone's education.

But there is another reason for studying astronomy. It allows us to probe the laws of nature under far more extreme temperatures, pressures and energies than can be achieved in laboratories here on Earth. It also allows us to study the fundamental force of gravity, and how it relates to the nature of space and time.

This is undoubtedly the Golden Age of astronomy. With the advent of the space age, new windows to the Universe have been opened. With giant observatories orbiting high above the Earth's atmosphere, one can now explore the Universe at a wide range of wavelengths: radio waves, millimetre waves, infrared radiation, visible radiation, ultraviolet radiation, X-rays and gamma rays. This has enabled astronomers to make unprecedented progress pertaining to a variety of questions: the nature of the stars and their life history; the formation of planets; the birth and death of the stars; the graveyard of stars—white dwarfs, neutron stars and black holes; galaxies; quasars; and the Universe at large.

This series of monographs entitled *The Present Revolution in Astronomy* is very timely for it aims to survey the contemporary scene at an introductory level.

Dr. G. Srinivasan, the author of this series of books, is an internationally acclaimed leader in this enterprise. In particular, he has studied neutron stars, which manifest an astonishing range of 'extreme' physics. Readers of these splendid and accessible books will find Dr. Srinivasan to be a clear and enthusiastic guide to the wonders and mysteries of the cosmos. We should all be grateful to him.

Cambridge Martin J. Rees

Preface to the Edition Published by Universities Press

The year 2009 was celebrated as the *International Year of Astronomy*. This was to commemorate the 400th anniversary of Galileo's pioneering observations with a telescope, observations that revolutionised man's perception of the heavenly bodies.

Four centuries later, we are in the midst of another golden era in astronomy. The advent of the space age has opened new windows to the Universe, resulting in spectacular discoveries and unprecedented progress in our understanding of the nature of celestial objects. At the same time, many new and outstanding questions have emerged. Indeed, there are clear indications that the resolution of some of these puzzles may require a major revision of fundamental physics itself. A deep connection between the *microcosm* and the *macrocosm* is becoming apparent.

This series of monographs entitled *The Present Revolution in Astronomy* is intended to convey the excitement of contemporary astronomy. The inspiration for writing these monographs was the enthusiastic response of the students who attended an intercollegiate course I taught for 5 years at St. Joseph's College in Bangalore. This course was not part of the regular academic curriculum, and was open to interested students and teachers from all the colleges in the city. Interestingly, more than half of the students in each batch were students of engineering, rather than pure science. And yet, they were fascinated by the lure of astronomy. Although the underlying theme of the course was *The Present Revolution in Astronomy*, my idea was to use astronomy as a *Trojan horse* to get the young students excited about the challenges that await them in the world of physics/astronomy, engineering and technology. It was the unanimous view of these students that I should develop these lectures into a series of books.

There is a second reason why I thought it would be worthwhile to write these books. Historically, astronomy has always had a great appeal among the general public. It is even more so today. The commissioning of new telescopes, and the discoveries made with them receive wide publicity in the print as well as the electronic media. Space Agencies like NASA, as well as leading astronomical observatories, have impressive Public Outreach programmes. And yet, here in India, hardly any of the universities offer astronomy as one of the subjects in the undergraduate curriculum. As a result of the lack of familiarity with the subject, very few students opt for a career in astronomy even though there are several truly world class observing facilities in India. This series of books is intended to partly remedy this lacuna.

Now, a few words about the scope of these monographs and the style in which they are written. My primary objective is to introduce the reader, young and not so young (!), to the presently unfolding revolution in astronomy. We shall discuss the recent developments concerning a wide variety of topics: *the nature of the stars and their life history; the birth and death of the stars; the graveyard of stars—white dwarfs, neutron stars and black holes; galaxies; quasars; and the Universe at large*.

The monographs are not intended to be 'textbooks' in astronomy. Textbooks have to develop the subject in a pedagogical manner, dwell on the experimental methods and phenomenology, develop the mathematical aspects of the theory in a systematic manner, include problems and exercises, etc. While all these are needed to learn a subject seriously, conventional textbooks have a serious handicap. Introductory books 'begin at the beginning' and seldom convey the excitement surrounding contemporary developments. They tend to focus on questions that have been resolved, rather than highlight what is not known. In contrast, this series of books is intended to serve a different purpose. I hope they will give the reader an introduction to the recent developments, as well as highlight the outstanding and unsolved questions. I believe that a young reader would be more interested in the unsolved puzzles, for that is where the challenges lie.

The books have a very different flavour compared to the traditional astronomy books. For example, they do not discuss topics such as measurement of distances to celestial objects, determination of their masses, luminosities, etc. Nor do they dwell on coordinate systems to define their positions in the sky, the classification of their spectra, etc. While all these are 'bread and butter' issues, it is my view that a reader would learn these topics at a later stage in the normal course if he or she decides to become a practising astronomer. The emphasis in this series of monographs will be on physics, and for the following reason.

Among the many great discoveries made by Isaac Newton, perhaps the most profound was his assertion that *the Laws of Nature have universal validity*. In other words, the laws of physics that govern phenomena on Earth apply everywhere in the Universe. Today, we take this assertion by Newton as an axiom. Indeed, during the past couple of centuries, several seminal inputs to laboratory physics have come from astronomical observations. The discovery of the law of gravitation, emission and absorption lines in the spectrum of the atoms, the discovery of Helium, the first verification of the predictions of the Special Theory of Relativity and the General Theory of Relativity are some of the more important examples. This is not surprising. The range of densities, temperatures and pressure that are obtained in celestial bodies are staggering compared to what one encounters on Earth. For example, the densities range from 1 atom/cm^3 to 10^{37} atoms/cm^3, and the temperatures range from 3 kelvin to 100 million kelvin—conditions that are hard for us to comprehend. Consequently, one encounters many new and exotic physical phenomena in celestial objects. Indeed, a few decades ago one would have said that *Astronomy is the home of physics*. Today, however, it would be more appropriate to say that *Physics is the home of astronomy*. We shall see the reason for this paradigm shift as we progress in this series. Therefore, we shall concentrate on the physics of the celestial bodies—their nature, their stability, their central engines, their radiation mechanisms, etc.

Having stated the objective of this series of books, I must add that I do not assume any astronomical background from the reader. A knowledge of physics at, say, the *Halliday and Resnick* level would be quite adequate to get started. We shall develop the rest of the background as we go along. To meet the stated objectives, I shall often be required to sacrifice rigour in the arguments in favour of simple analogies and qualitative arguments. And I shall do so without any apologies! I shall consider my efforts worthwhile if these books manage to convey the excitement of contemporary astronomy. As for the younger readers, I do hope that these books will arouse their interest sufficiently enough for them to want to pursue the topics further by going to more learned books.

When I was young, I had the pleasure and privilege to read the marvellous books by great masters like Sir Arthur Eddington, Sir James Jeans and George Gamow, books in which they explained the developments in physics and astronomy in the early part of the last century. There are several recent books, written in a similar vein, by leading physicists and astrophysicists, of the present epoch. And then there is the 'Internet'! This series of monographs represents my very humble efforts in the same spirit.

This Volume

In the first volume in this series, entitled *What Are the Stars?*, I discussed the nature of the stars, their stability and the origin of the energy they radiate. One of the fascinating things about stars is that they evolve as they age. This evolution is different for stars of different masses. How stars end their lives when their supply of energy is exhausted also depends on their mass. This volume is devoted to a discussion of the evolution of stars and their ultimate fate. Historically speaking, astronomers first worried about the ultimate fate of the stars, even before the details of their evolution became clear.

I have divided this volume into two parts. The Part I is an account of the remarkable predictions made during the 1920s and 1930s concerning the ultimate fate of the stars. Since much of this development hinged on the emerging quantum physics, I have given a detailed introduction to the relevant physics. These topics will be useful to you should you decide to pursue studies in condensed matter physics, nuclear physics, astrophysics, etc.

Part II is a summary of the life history of stars. This discussion is divided into three parts: low-mass stars like our Sun, intermediate-mass stars, and massive stars.

As you read this volume, you will discover that much of contemporary astrophysics has been built on the foundations erected by Subrahmanyan Chandrasekhar in the 1930s. Since this volume has been written during his birth centenary, I have included in it a brief biographical sketch of Chandrasekhar.

Acknowledgments

The idea of this series was first suggested by the students who attended the intercollegiate course on astronomy and astrophysics that I taught for a number of years at St. Joseph's College in Bangalore. This suggestion was strongly endorsed by Dr. P. Sreekumar of the ISRO Satellite Centre. The enthusiastic response of the student community to the series of books entitled *Vignettes in Physics*, written by Dr. G. Venkataraman, as well as Venkataraman's eloquent and sustained persuasion that I should write a similar series on contemporary astronomy, gave me the conviction I needed to undertake this task. A further impetus came when the Jawaharlal Nehru Memorial Fund bestowed on me the *Jawaharlal Nehru Fellowship* in 2007 to get started on this project. In 2009, the Nehru Centre in Mumbai was very kind to give me a Fellowship for 2 years to continue with the project. I am very grateful for both these Fellowships. I started out as a condensed matter physicist, but later wandered into astronomy! My first introduction to astronomy came from my father at an early age. The inspiration to pursue it and the attempt to popularise it, came first from my illustrious teacher Subrahmanyan Chandrasekhar (at the University of Chicago), and later from Profs. Martin Rees (Cambridge University), Ed van den Heuvel (University of Amsterdam) and V. Radhakrishnan (at the Raman Research Institute, Bangalore). I am most grateful to them for having inspired me. I would like to express my special thanks to NASA, ESA, and the international astronomical fraternity for the many wonderful images reproduced in these volumes.

G. Srinivasan

Contents

Part II The Life History of Stars: A Modern Perspective

About the Author

Dr. G. Srinivasan began his career as a solid-state physicist and later switched to astrophysics. After his Ph.D. at the University of Chicago, he worked at the IBM Research Laboratory, Zurich, Switzerland, Chalmers University of Technology, Goteborg, Sweden, Cavendish Laboratory, University of Cambridge and Raman Research Institute, Bangalore. He is a Past President of the Astronomical Society of India as well as the Division of Space and High Energy Astrophysics of the International Astronomical Union. He is a Fellow of the Indian Academy of Sciences and a former Jawaharlal Nehru Fellow.

Part I
A Historical Perspective

Chapter 1
What Are the Stars?

Globes of Gas

This chapter is intended to be a brief recap of the relevant parts of the first volume in this series entitled ***What Are the Stars?***

The major breakthrough in our understanding the nature of stars came with the discovery by Fraunhofer in 1817. He demonstrated that the spectrum of sunlight contained *dark lines*. In the mid 1850s, Kirchoff and Bunsen demonstrated in laboratory experiments that such dark lines could be produced in the spectrum of light from an opaque body by *passing the light through transparent substances.* This led Kirchoff to formulate his comprehensive theory of radiation. It became clear from these investigations that *the outer layers of the Sun and the stars were gaseous, with a composition similar to what we find on Earth.* Thus, the picture emerged that the stars are globes of gas, held together by their own gravity.

The beginning of our understanding of the true nature of the stars can be traced back to the second half of the nineteenth century. **J. Homer Lane** was the first person to investigate the details of the temperature distribution within a star. In 1870, he published a seminal paper in *American Journal of Science and Arts,* entitled, 'On the theoretical temperature of the Sun, under the hypothesis of a gaseous mass maintaining its volume by its internal heat, and depending on the laws of gases as known to terrestrial experiment'. Put simply, in this work Lane assumed that stellar matter behaved as an *ideal gas* and obeyed Boyle's law, as terrestrial gases do. The basic idea was that the inward pull due to self-gravity was balanced by the pressure of the gas, as indicated in Fig. 1.1.

But how do we know that the interior of the Sun is hot? A very simple argument tells us that the mean temperature inside the Sun must be of the order of a few million degrees kelvin.

G. Srinivasan, *Life and Death of the Stars*, Undergraduate Lecture Notes in Physics,
DOI: 10.1007/978-3-642-45384-7_1, © Springer-Verlag Berlin Heidelberg 2014

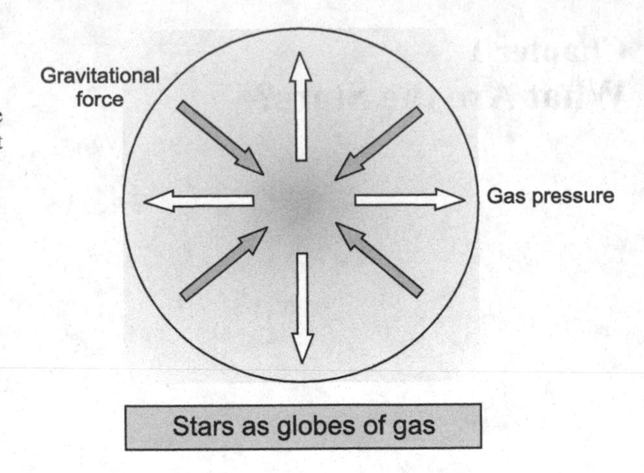

Fig. 1.1 A star is stable because the inward-directed force due to self-gravity is opposed, and balanced, by the pressure of the gas. The weight of the column of gas above any point in the star must be countered by the pressure of the gas. This condition must be satisfied at every point in the star. Otherwise the star will not be in mechanical or *hydrostatic equilibrium*

The Temperature of the Sun

To see this, let us invoke the famous *virial theorem*. This theorem is of very general validity and is applicable as long as the system under consideration is statistically stable. The theorem states that in the steady state *the total energy of the system is equal to one-half its potential energy*. This powerful theorem can be invoked to estimate the average temperature of the Sun. In this case, the total energy is the sum of the stored *thermal energy* in the Sun and the *gravitational potential energy* of the Sun due to self-attraction. According to the virial theorem,

$$\boxed{\text{Thermal Energy} + \text{Grav. Potential Energy} = \frac{1}{2}\text{Grav. Potential Energy.}}$$

Therefore,

$$\boxed{\text{Thermal Energy} = -\frac{1}{2}\text{Grav. Potential Energy.}} \tag{1.1}$$

(Notice the *minus sign* on the right-hand side. Remember that the gravitational potential energy is negative. Therefore the minus sign is needed to make the right-hand side positive). The gravitational potential energy of a sphere of mass M and radius R is $\sim -GM^2/R$. The thermal energy of the Sun is just the sum of the kinetic energy of the constituent particles. Let T be the average temperature of the Sun. We know from the kinetic theory of gases that the *average energy of the particles* is $\frac{3}{2}k_BT$. If N is the total number of independent particles, then the total thermal energy is $\frac{3}{2}Nk_BT$. Thus, according to the virial theorem (1.1),

$$\frac{3}{2}Nk_BT = \frac{1}{2}\frac{GM^2}{R}. \tag{1.2}$$

We know the mass and radius of the Sun. We can estimate the number of particles by assuming some chemical composition. The above equation can then be solved for the average temperature. This yields a value of *10 million kelvin*. (Take a few minutes to verify this. Assume for simplicity that the Sun is made solely of hydrogen. Since you know the mass of the Sun, you can estimate the number of atoms in the Sun.) I hope you are astonished by the power of the virial theorem, which enabled us to make this estimate. *Sitting here on Earth, we can say with considerable confidence that the average temperature of the Sun must be ten million kelvin!* We only needed to know the mass and radius of the Sun to be able to make this estimate.

Hydrostatic Equilibrium

Next, let us set up the equation for the mechanical stability of the star. Consider an imaginary concentric spherical surface of radius r inside the star, as shown in Fig. 1.2. Let us place on this surface a small cylinder, whose axis points along the outward radius at that point. The cross-section of this cylinder is of unit area of the base and length dr and it contains stellar material. The density of this stellar material is $\rho(r)$, which is the value of density obtained at the distance r from the centre. The gravitational force on that cylinder would be due to the mass interior to the imaginary surface. Let us call this mass $M(r)$.

As the area of cross-section of the cylinder is unity and its length is dr, the mass of the infinitesimal cylinder is given by $\rho(r)dr$. The force of attraction between $M(r)$ and $\rho(r)dr$ is

$$\frac{GM(r)\rho(r)dr}{r^2}. \tag{1.3}$$

As you know, in Newton's law the contribution to the force from the mass *exterior* to the surface cancels out. If you know some calculus, I urge you try and prove this. You will find it illuminating. The gravitational force on this infinitesimal cylinder has to be balanced by the pressure differential on it.

This is just the *difference* between pressure measured at the two surfaces of the cylinder at a distance from the centre equal to r and $r + dr$, respectively. Let us denote this by dP. This pressure difference dP represents the force $-dP$ acting on the cylinder in the direction of increasing r. Thus the equation for the equilibrium of the unit cylinder is

$$dP = -\frac{GM(r)\rho(r)dr}{r^2}.$$

One can rearrange this as

$$\boxed{\frac{dP}{dr} = -\frac{GM(r)\rho(r)}{r^2}}. \tag{1.4}$$

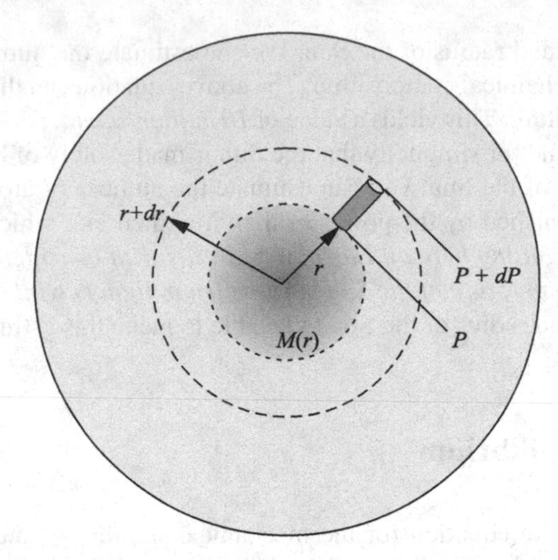

Fig. 1.2 Consider an infinitesimal cylinder at a distance **r** from the centre of unit cross-sectional area and of height **dr**. The gravitational force acting on it will arise from the mass **M(r)** of material interior to the spherical shell on which it lies. This has to be balanced by the difference in pressure **dP** which represents a force −**dP** in the direction of increasing **r** (pointing outward from the centre). This is the condition for hydrostatic equilibrium of the star, *and must be satisfied at every point in the star*

The above equation is known as the *equation of hydrostatic equilibrium*. For a star to be mechanically stable, this equation has to be satisfied *at every point in the star*. Otherwise, as a distinguished astronomer said, 'the punishment would be swift'. Any violation of this condition of *hydrostatic equilibrium* would result in motions within the star. For example, the material within our sample unit cylinder would either sink or float up due to buoyancy.

Radiative Equilibrium

In Lane's theory the pressure on the left-hand side of (1.4) is the pressure of an ideal gas: $p_G = nk_BT$, where n is the number density of particles. He further assumed that the internal heat is transported outwards by *convection*, very much like what happens in our atmosphere. Around 1920, **Sir Arthur Eddington** at Cambridge University introduced the idea of *radiative equilibrium*. In this scenario, *the heat flowing outward is transported by radiation*, rather than convection or conduction. This flux of radiation flowing from the interior towards the surface will exert pressure on the stellar material. Eddington's point was that the pressure that supports against gravity is the *sum* of gas pressure and radiation pressure.

You will recall that the radiation has momentum E/c, where E is the energy and c is the velocity of light (in the quantum picture, the momentum of a photon is $h\nu/c$,

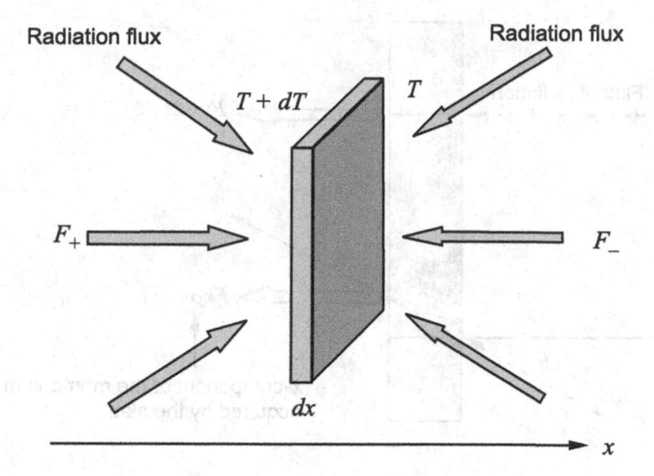

Fig. 1.3 Radiation passes through a slab of stellar material of unit area and thickness dx from both sides. Let the temperature on the two faces be $T + dT$ and T, respectively. Consequently, the flux of radiation and the pressure of radiation on one side will be more than on the other. There will be a resultant pressure $-dp_R$ in the direction of the temperature gradient

where h stands for Planck's constant and ν is the frequency of the photon). Since momentum is associated with radiation, it must exert pressure, just as gas particles do. Let us consider a special kind of radiation known as *black body radiation*. This is just radiation in an enclosure with absorbing walls maintained at a temperature T. Given enough time, the radiation in the cavity will come to thermal equilibrium with the walls. It will be isotropic as regards the direction of flow and will be characterized uniquely by the temperature of the walls of the cavity. An important result from the nineteenth century is that *the energy density of radiation in the cavity is proportional to the fourth power of the absolute temperature:*

$$E = aT^4, \tag{1.5}$$

where a is a universal constant known as *Stefan's constant*. The above relation is known as *Stefan's law*. The pressure exerted by this radiation is

$$\boxed{p_R = \frac{1}{3}aT^4} \tag{1.6}$$

Let us now understand the principle of radiative equilibrium. Pick any radial direction in the star and call it the x axis, and let the positive direction of this axis be along the temperature gradient. Consider a slab of stellar material of thickness dx and area equal to one square centimetre held normal to the x axis (see Fig. 1.3).

Let the temperature of the two faces of the slab be T and $T + dT$, respectively. Since pressure is force per unit area, the forces exerted by radiation on the two faces are $+p_R$ and $-(p_R + dp_R)$. The resultant force in the direction of the temperature

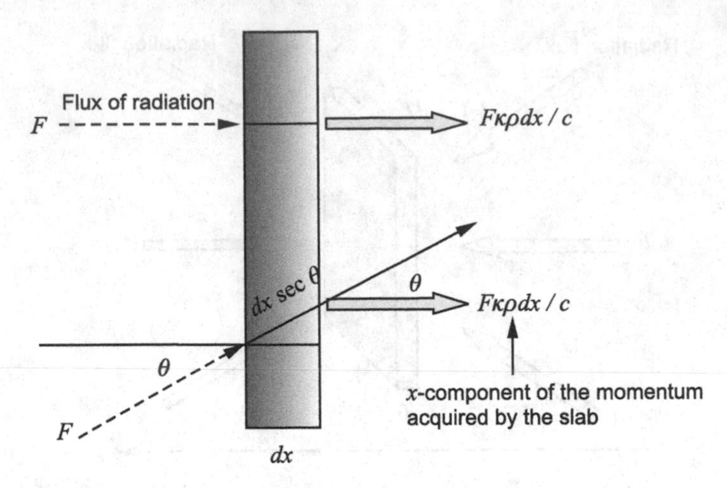

Fig. 1.4 The fraction of the radiation flux F absorbed by the slab will be equal to the flux multiplied by the mass absorption coefficient *per unit mass* multiplied by the *volume* of the slab, that is, $F\kappa\rho dx$. Therefore the x-component of the momentum acquired by the slab per unit time will be $F\kappa\rho dx/c$. For the slab to be in equilibrium, this x-component of the momentum absorbed from the radiation must be equal to $-dp_R$

gradient is $-dp_R$. We have adopted the convention that the force is positive if it is in the direction of gravity and negative if directed outwards.

This resultant force imparts momentum to the slab. For the slab to be in equilibrium it must utilize this momentum in some fashion; otherwise, the slab will be set in motion. What the material of the slab does is to *absorb* this momentum and *use it to supplement the gas pressure* in its attempt to support itself against gravity.

Next we have to calculate the x-component of the momentum absorbed by the material in the slab. Let us first introduce the mass absorption coefficient κ. This is the coefficient of absorption *per gram* of matter. Let F be the flux of radiation incident on the slab (measured in *ergs per square centimetre per second*). The fraction of the flux absorbed by the slab will be $F\kappa\rho dx$, where ρ is the density of matter in the slab. Since the area of the slab is unity and its thickness is dx, the mass of the slab is just ρdx (see Fig. 1.4). The x-component of the momentum absorbed by the material per unit time is

$$F\kappa\rho dx/c, \tag{1.7}$$

where c is the velocity of light (Interestingly, the above result holds even if the radiant flux is incident obliquely. If the angle of incidence is θ then the distance travelled by it through the slab is increased to $dx \sec\theta$. So the energy absorbed in the slab increases by $\sec\theta$. But the x-component of the momentum absorbed remains unchanged because to obtain the x-component, we have to multiply the above by $\cos\theta$ which cancels $\sec\theta$).

Finally, we want to calculate the net momentum absorbed by the slab per unit time. Remember that radiation is incident on the slab from both sides. Let us denote the flux from the left (outward flowing) by F_+ and the flux from the right by F_-. The *net outward flux* is given by

$$F = F_+ - F_-, \tag{1.8}$$

and the net positive momentum gained by the slab is $F\kappa\rho dx/c$. Earlier we said that for the slab to be in radiative equilibrium, the momentum gained per second by the slab must be fully absorbed by the matter contained in it. Hence,

$$-dp_R = F\kappa\rho dx/c,$$

or

$$\boxed{F = -\frac{c}{\kappa\rho}\frac{dp_R}{dr}.} \tag{1.9}$$

(We have replaced x by the radial co-ordinate, r). Substituting for the radiation pressure from Stefan's law, $p_R = \frac{1}{3}aT^4$, we get for the net *outward flux:*

$$\boxed{F = -\frac{ac}{3\kappa\rho}\frac{dT^4}{dr}.} \tag{1.10}$$

This is the famous result obtained first by Eddington. It says that *net flux of radiation* is proportional to the *pressure gradient* and inversely proportional to the *opacity* of the stellar matter (Eddington called $\kappa\rho$ the *obstructive power of the material screen* through which the radiation is forced).

Eddington's Theory of Stars

Eddington constructed a theory of stars based on the above principle of radiative equilibrium. According to him, the pressure that balances gravity in Eq. (1.4) is the sum of gas pressure and radiation pressure,

$$P = p_G + p_R, \tag{1.11}$$

where

$$p_G = nk_BT = \frac{\rho k_B T}{\mu m_p}, \quad p_R = \frac{1}{3}aT^4. \tag{1.12}$$

In Eq. (1.12) we have expressed Boyle's law in terms of the mass density ρ. Since we shall be doing this in other contexts also, let us understand how this is done. If our gas consisted of only one species of particles, then the number density, n, and the mass density, ρ, are related by

$$\text{number density} = \frac{\text{mass density}}{\text{mass of the particle}}.$$

The stellar plasma, however, consists of more than one species of particles; it consists of electrons and nuclei of different elements. Therefore the correct thing to do would be to divide the mass density by the *average mass* of the independent particles:

$$\text{number density} = \frac{\text{mass density}}{\text{average mass of the particles}}.$$

To define the average mass of the particles one needs to know the chemical composition of the plasma. It is customary to introduce the notion of the *mean molecular weight* μ in defining the relation between the *number density* of independent particles, n, and the *mass density*, ρ,

$$n = \frac{\rho}{\mu m_p}, \tag{1.13}$$

where m_p is the mass of the proton (since the mass of the neutron is very nearly the same as that of the electron, we shall not distinguish between the two). The terminology *molecular weight* is borrowed from chemistry and is a misnomer here. In the present context, the term *molecule* really refers to the independent particles of our gas, nuclei of different species and the electrons. It should be clear from the above equation that μm_p is defined as *the average mass of the independent particles of the gas.*

The Mass–Luminosity Relation

One of the most spectacular predictions of Eddington's theory concerns the relation between the mass and the luminosity of a star. The theory predicts that the luminosity of a star is proportional to the cube of the mass,

$$\boxed{L \propto M^3.} \tag{1.14}$$

This is a remarkable result. Notice that the radius of the star does not enter! One would think that given a star of a certain mass, the luminosity it generates should depend on its radius. After all, the internal temperature should be determined by the radius—common sense tells us that smaller the star, the hotter it would be—and that, in turn, should determine the rate of energy generation.

But the radius does not enter the expression for the luminosity. It is almost as though the star knows the radius it must attain. Well, it does! The principle of radiative equilibrium *dictates* to the star the luminosity it is allowed to generate, and that luminosity is determined only by its *mass* and the *opacity* [notice that the opacity or the mass absorption coefficient enters Eq. (1.10) for the net outward flux of radiation].

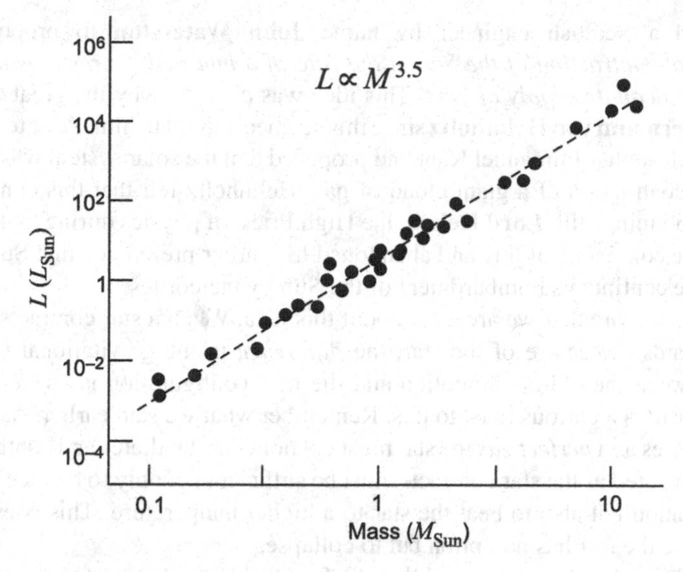

Fig. 1.5 A log–log plot of the mass–luminosity relation using recent data. As shown, an exponent of 3.5 fits the data very well. (From Encyclopedia.com: http://www.encyclopedia.com/doc/1E1-masslumi.html, with the courtesy of the author.)

Given the opacity of the stellar material, the star will adjust itself to that combination of RT such that *the energy generated per unit time precisely compensates for the heat energy lost from the surface per unit time.* If the star were to generate more luminosity, given that the rate at which the energy can diffuse outwards is determined by the opacity, there will be a buildup of energy in the interior and the condition of radiative equilibrium will be violated.

The prediction that $L \propto M^3$ was in excellent agreement with observations. More recent data are shown in Fig. 1.5. As will be seen, $L \propto M^{3.5}$ gives an excellent fit to this data set. This slope is very nearly what Eddington's theory predicts.

Why Do the Stars Shine?

A much debated question towards the end of the nineteenth century concerned the source of energy radiated by the Sun. Lane's theory predicted a curious behaviour for the stars. As the star radiates energy, the internal temperature must decrease (since the internal energy is being radiated away). This will disturb the delicate balance between the gravitational force and the pressure force. Consequently, gravity will gain an upper hand, and the star will have no option but to contract. But this will compress the gas and make it hotter. *So we have the curious behaviour that as the star radiates energy, it will get hotter!* The stellar material has *negative specific heat*, in violation of the laws of thermodynamics.

This led a Scottish engineer by name **John Waterston** to propose that *gravitational contraction of the Sun at the rate of a hundred metres a year would provide an adequate supply of heat*. This idea was picked up by the great German physicist **Hermann von Helmholtz** since this seemed natural to him. Prior to this, the German philosopher Immanuel Kant had proposed that the solar system was formed due to the contraction of a giant cloud of gas. Helmholtz felt that this contraction must be continuing still. **Lord Kelvin**, the High Priest of physics during that period, also became convinced of this and abandoned his earlier preference that Sun's heat is due to the continuous bombardment of the Sun by meteorites.

Let us make sure that we are clear about this idea. When a star contracts, matter moves towards the centre of the star; the *difference* in the gravitational potential energy between the old configuration and the new configuration is converted into heat. But there is a curious twist to this. Remember what we said earlier. As long as the gas behaves as a *perfect gas* the star must get hotter as it radiates and contracts. So the heat generated as the star contracts must be sufficient not only to replace the heat lost as radiation but also to heat the star to a higher temperature. This is essential, for otherwise the star has no option but to collapse.

Helmholtz and Kelvin estimated that the Sun had been shining for about twenty million years, and will continue to shine for another twenty million years or so. Let us see how one may estimate this timescale. Recall our discussion of the *virial theorem*: When a star contracts, *only one-half of the gravitational potential energy released is available for radiation.* The other half is stored as thermal energy as explained in Eq. (1.1). The gravitational potential energy of the Sun is $\sim -2GM_\odot^2/R_\odot$. If we divide one-half of this by the rate at which the Sun has been radiating, then we can get an estimate of how long the Sun has been shining.

$$t \sim \frac{-\frac{1}{2}\text{gravitational potential energy}}{\text{Luminosity}} \sim \frac{GM_\odot^2/R_\odot}{L_\odot}. \tag{1.15}$$

The present rate at which the Sun is losing energy is the *Luminosity* of the Sun [$L_\odot = 4 \times 10^{33}$ erg s^{-1}]. Inserting the values for the mass and radius of the Sun, we come to the conclusion that if the Sun had been radiating at the present luminosity, then it could have done so only for about 20 million years [Convince yourself of this by substituting the values]. This seemed a comfortably long time for Lord Kelvin. Even though the geologists were convinced (even at that time) that the Earth was older than 20 million years, Lord Kelvin was not bothered. He used his status to tell the geologists to confine themselves to this timescale!

The discovery of radioactivity was the last nail in the coffin for the contraction hypothesis. Using modern techniques, geologists were able to determine the age of the older rocks, and this turned out to be *more than a billion years.* Now, if the earth itself is several billion years old, the Sun must be even older than this. So, it was back to square one, regarding explanations about the source of energy in the Sun and the stars. But one could say this: if an external source of energy (such as meteorites), as well as gravitational contraction, are ruled out then *the star must contain some hidden source of energy which enables it to shine for billions of years.*

What is this hidden source of energy? Sir Arthur S. Eddington provided the breakthrough. Addressing the *British Association* in Cardiff on 24 August 1920, Eddington argued that only *subatomic energy* is available in unlimited quantity. What inspired Eddington to make this remarkable assertion was the discovery that had just then been made by F. W. Aston, one of Rutherford's students at the Cavendish Laboratory in Cambridge. Using the *mass spectrograph* that he had just invented, Aston was able to measure the masses of the atoms. One of the discoveries he made was that *the mass of four hydrogen nuclei was greater than the mass of one helium nucleus.* Eddington's idea was that if four protons fuse to produce a helium nucleus then this *mass deficit* would be converted into energy according to Einstein's formula:

$$E = \Delta M c^2.$$

Let us examine this closely. The mass of four protons is $4 \times 1.0081 m_u$ (atomic mass units), while the measured mass of the ^{4} He nucleus is $4.0039 m_u$. This means that a mass of $2.85 \times 10^{-2} m_u$ has *disappeared* for every helium nucleus produced if, indeed, the helium nucleus was produced by fusing four protons. This is roughly 0.7 % of the original mass of hydrogen, and corresponds to energy measuring about 26.5 MeV. Another way to say this is the following. If mass M of hydrogen is converted into helium, then the energy released is 0.007 Mc^2 (Think of James Bond to remember this formula!). The mass of the Sun is 2×10^{33} g, most of it hydrogen. By converting most of it to helium, it can generate $\sim 10^{52}$ erg of energy. The rate at which it radiates this energy (its luminosity) is 4×10^{33} erg s^{-1}. Therefore, the Sun can easily shine for 10^{11} years by tapping this source of *subatomic energy.*

$$t_{\text{nuclear}} \sim \frac{0.007 M_\odot c^2}{L_\odot},$$

$$t_{\text{nuclear}} \sim \frac{0.007 \times 2 \times 10^{33} \times 10^{21}\text{erg}}{4 \times 10^{33}\text{erg s}^{-1}} \sim 10^{11}\text{years}. \tag{1.16}$$

Can Stars Find Peace?

How long a star will shine will depend, of course, on its mass. The above estimate is for a star of one solar mass. From Fig. 1.5, we saw that the luminosity of a star is roughly proportional to $M^{3.5}$. Hence the nuclear timescale will be roughly proportional to $M^{-2.5}$. In other words, a massive star will have a shorter lifetime than a star with low mass. Although the massive star has more fuel, it spends it more furiously!

What will happen to a star when its supply of nuclear energy is exhausted? Will it collapse to a point and disappear from this Universe, or, is there a new twist to the story?

This book is devoted to a discussion of this question.

Chapter 2
Stars in Their Youth

The Hertzsprung–Russell Diagram

Perhaps the most important diagram in stellar astronomy is what is known as the Hertzsprung–Russell diagram (H–R diagram). It is a plot of the luminosity of a star versus its surface temperature (also known as the effective temperature). Most stars you see in the sky when plotted in this diagram fall into a diagonal band known as the *main sequence*. What is shown in Fig. 2.1 is a theoretical H–R diagram.

An important property of all stars that fall into this band is that they may be regarded as chemically homogeneous, and are converting hydrogen to helium in their cores. In a real sense, all the stars along this sequence have 'recently' formed out of the interstellar gas. For this reason, this main sequence is often referred to as the *zero-age main sequence* (ZAMS). This phase, during which hydrogen is being fused into helium, has such a long duration that most stars visible in the sky are likely to be in this phase (since stars spend most of their lives in this phase, the probability of catching them in this phase will obviously be the greatest). If all the stars found in the main sequence are chemically homogeneous and are converting hydrogen into helium in their core, one may ask what distinguishes them. The most important factor that determines the location of a star *within* the main sequence is the *mass of the star*. Notice that in Fig. 2.1 the more massive stars have a higher luminosity, as one would expect from Eddington's theory.

The solid lines in Fig. 2.1 are loci of constant radii. Thus a $1 M_\odot$ *zero-age star* has a radius very nearly equal to $1 R_\odot$. You may think this is a bit shady. Would one not expect a $1 M_\odot$ star to have a radius precisely equal to $1 R_\odot$? Well, the *present* radius of the Sun is what we call $1 R_\odot$. The $1 M_\odot$ star in Fig. 2.1 is a *zero-age* star. The Sun descended on the main sequence nearly 5 billion years ago and its radius had changed somewhat during this period. Similarly, a $10 M_\odot$ zero-age star has a radius somewhat less than $10 R_\odot$. This would suggest that the radius is roughly proportional to the mass. More careful consideration shows that in the lower part of the main sequence

G. Srinivasan, *Life and Death of the Stars*, Undergraduate Lecture Notes in Physics,
DOI: 10.1007/978-3-642-45384-7_2,

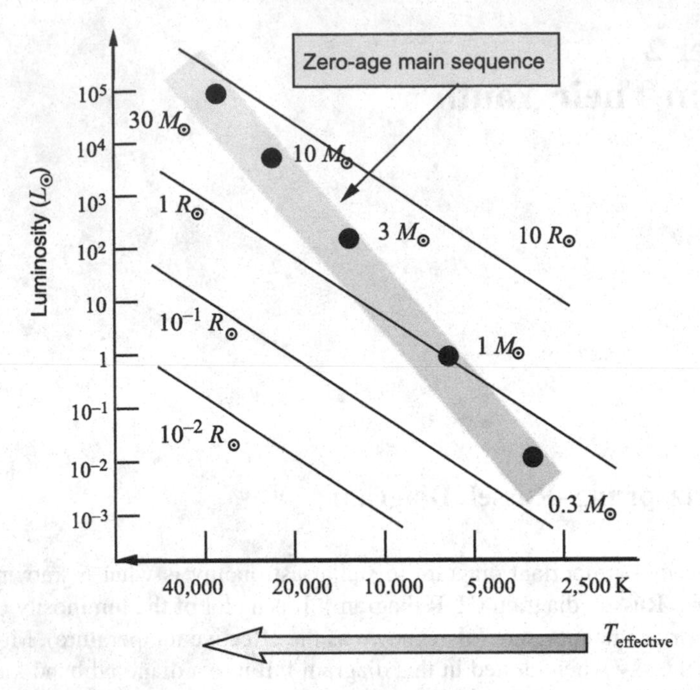

Fig. 2.1 The diagonal band shows the location of the main sequence in the theoretical Hertzsprung–Russell diagram. This famous diagram is a plot of the luminosity versus the surface temperature (also referred to as the *effective temperature*) of stars. *Notice that the effective temperature increases from right to left!* The points indicate the theoretical location of stars of various masses when they begin their lives

$$R \propto M, \tag{2.1}$$

while in the upper part of the main sequence (that is, for the more massive stars)

$$R \propto M^{0.6} \tag{2.2}$$

is a better approximation. The above relation between the radius and the mass, taken together with the theoretical mass-luminosity relation will tell us how the surface temperature will depend on the mass. We saw in Fig. 1.5 that $M^{3.5}$ gives a reasonable fit to the observational data on luminosity. But this is over the whole range of masses. In the range 1–10 $M_\odot$ the data is better fit by an exponent 4, that is

$$L \propto M^4. \tag{2.3}$$

Using $R \propto M$ and $L \propto M^4$, together with $L = (4\pi R^2)\sigma T_{\text{eff}}^4$, we get

$$T_{\text{eff}} \propto M^{1/2} \tag{2.4}$$

on the main sequence. We know that the effective temperature of the Sun is 5,800 K. Using this to deduce the proportionality constant in (2.4) we can deduce that a $10M_{\odot}$ star will have a surface temperature of about 20,000 K. To put it differently, while the Sun is a *yellow star*, a $10M_{\odot}$ star will be a *blue star*. Remember Wien's displacement law? The wavelength at which the black body spectrum has a maximum depends upon the surface temperature of the black body.

The moral of the story is this: *High-mass main sequence stars are more luminous and intrinsically bluer than low-mass main sequence stars.*

Energy Generation in the Main Sequence

As mentioned above, the most distinguishing feature of the stars on the main sequence is that they are converting hydrogen into helium in their cores. In the Chap. 1 we outlined the extraordinary conjecture by Eddington. But it took nearly twenty years to work out the details. The first breakthrough in solving the problem of how stars liberate energy came in 1938 when **C. F. von Weizsäcker** discovered a nuclear cycle, now known as the *carbon–nitrogen–oxygen (CNO) cycle*, in which hydrogen nuclei could be fused using carbon as a catalyst. However, von Weizsäcker did not work out the rate at which energy could be produced in the stars using this CNO cycle or how this rate would depend on the temperature that obtains in the stars.

The credit for this must go to **Hans Bethe**, the acknowledged master of nuclear physics. In 1938, Bethe had just completed a set of three monumental review articles in nuclear physics. These were known as *Bethe's Bible.* The first textbooks in nuclear physics were published only several years after the end of World War II. Until then, physicists all over the world learnt their nuclear physics from these pedagogical and authoritative articles by Bethe. In the 1930s, physicists were not concerned with problems in astronomy. They were more interested in atomic and molecular spectra, and nuclear physics. It was George Gamow who sensitized physicists about the unsolved problems concerning stellar physics by convening a small conference in Washington, D.C. Hans Bethe and many of the leading physicists were at that conference. Within a few months of this, Hans Bethe had worked out, in great detail, the synthesis of helium in stars and published his results in a landmark paper entitled, *Energy production in stars* (1939). Bethe considered two processes. One of them has come to be known as the **p–p chain** in which one builds helium out of hydrogen. This is the process that is important for stars like the Sun, and stars of even lower mass. The other process is the **CNO cycle** discovered earlier by von Weizsäcker, and is the dominant process for stars more massive than the Sun.

We have discussed both these processes in detail in ***What Are the Stars?*** Here, we shall briefly recall the steps involved in these reactions by reproducing the relevant figures from there.

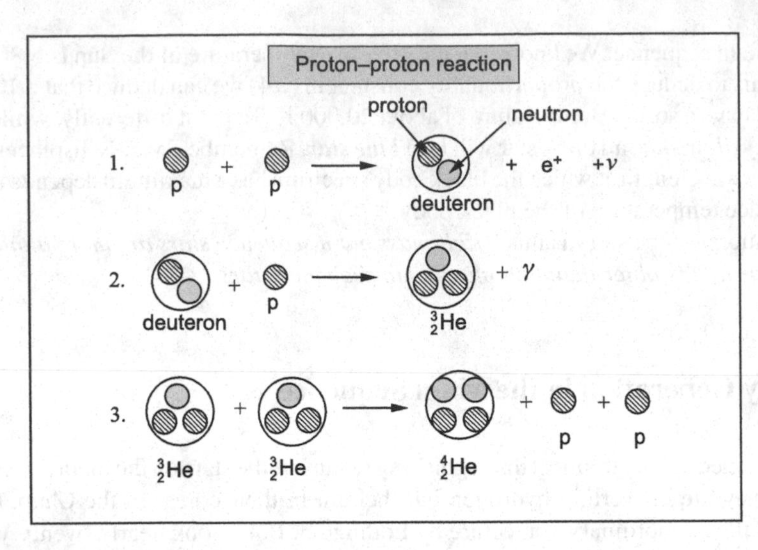

Fig. 2.2 The synthesis of protons into helium nucleus. This is the *main branch* of the p–p chain reaction, and accounts for 85 % of the energy generation. The remaining 15 % is through alternate branches, which we shall not discuss here

The p–p Chain Reaction

Figure 2.2 summarizes the main channel in the proton–proton chain.

The CNO Cycle

The other route for the synthesis of helium is the Carbon–Nitrogen–Oxygen cycle, first discovered by C. F. von Weizsäcker. The details of this were worked out by Hans Bethe in 1939. The CNO cycle requires the presence of some carbon, nitrogen or oxygen which act as catalysts in chemical reactions.

Here also, like in the p–p chain, four protons are fused into one helium nucleus, releasing roughly the same amount of energy as before (25 MeV per ^{4}He nucleus produced).

All nuclear reactions are sensitive to temperature. Fusion of nuclei is made possible by quantum mechanical tunnelling through the repulsive coulomb barrier. Given the like charges of the two colliding nuclei, the probability of such a tunnelling depends very sensitively on the kinetic energy of the particles. And this, in turn, depends upon the temperature of the stellar plasma (Figs. 2.3 and 2.4).

The p–p chain is the least temperature-sensitive of all the fusion reactions. The CNO cycle is much more sensitive to temperature. This has the consequence that the p–p chain dominates at lower central temperatures ($T_c < 15 \times 10^6$K). At higher

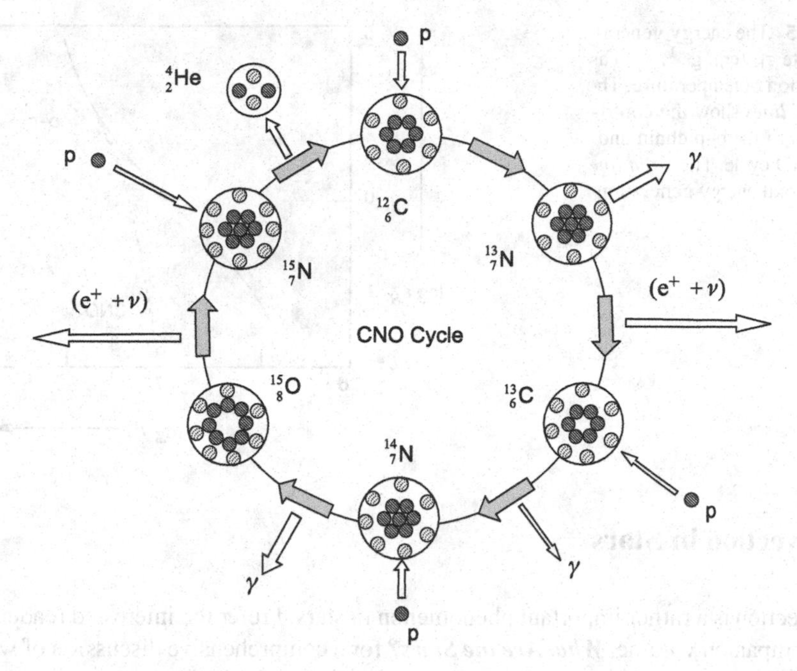

Fig. 2.3 A pictorial representation of the CNO cycle. The *bigger circles* represent the nuclei indicated. The small *hatched circles* are the protons and the *small circles* with *dots* are the neutrons

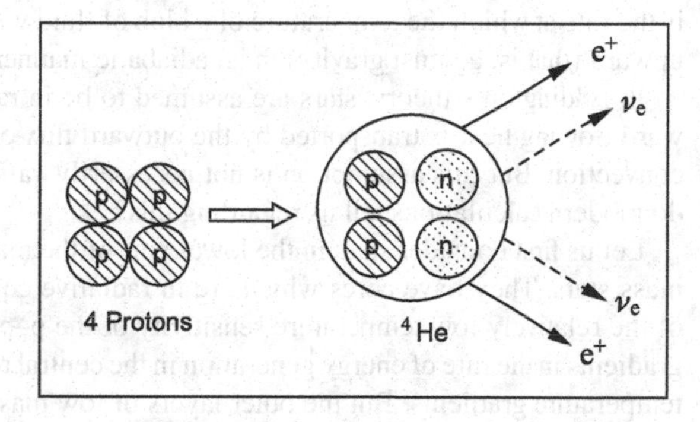

Fig. 2.4 In both the p–p reaction chain and the CNO cycle, four protons are involved in forming a helium nucleus. For every helium nucleus that is synthesized, *two positrons* and *two electron neutrinos* are emitted

central temperatures the CNO cycle dominates over the p–p chain. This is shown in Fig. 2.5.

Going back to the main sequence sketched in Fig. 2.1, for stars more massive than the Sun the CNO cycle is the main process of energy generation, while for stars less massive than the Sun the p–p chain is the main channel.

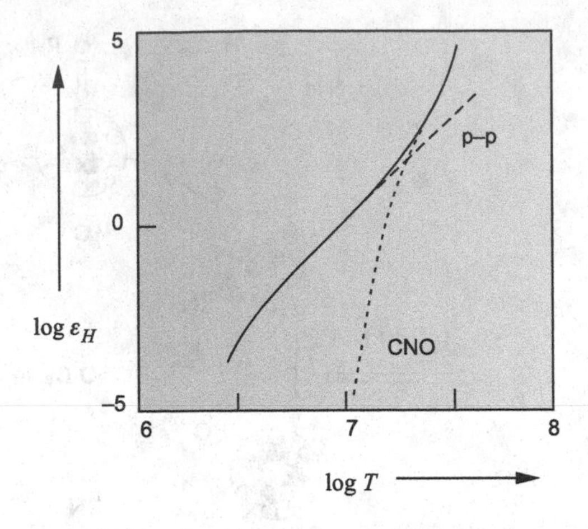

Fig. 2.5 The energy generation rate ε_H (erg. g^{-1}. s^{-1}) as a function of temperature. The *dashed lines* show the contributions of the p–p chain and the CNO cycle. The *solid line* is the total energy generation

Convection in Stars

Convection is a rather important phenomenon in stars. I refer the interested reader to the companion volume, ***What Are the Stars?*** for a comprehensive discussion of why convection occurs. Here, we shall merely state that *a fluid in a gravitational field will become unstable to the onset of convection if the temperature gradient exceeds a critical value*. This critical value is known as the *adiabatic temperature gradient*; this is the rate at which the temperature of a blob of fluid will decrease if it is transported upward (that is, against gravity) in an adiabatic manner.

In Eddington's theory, stars are assumed to be in radiative equilibrium; the outward flowing heat is transported by the outward flux of radiation itself. There is no convection. But this assumption is not necessarily valid everywhere in a star. What do modern calculations tell us regarding this?

Let us first consider stars in the lower part of the main sequence, namely the low mass stars. They have cores which are in radiative equilibrium. As a consequence of the relatively low temperature sensitivity of the p–p reactions, there are no steep gradients in the rate of energy generation in the central region and, therefore, no steep temperature gradients. But the outer layers of low mass stars tend to be convective. The outer layers of these stars tend to be cooler. This, in turn, increases the opacity of the outer regions. The presence of new species of ions, such as the *negative ion of hydrogen* (hydrogen atom with two electrons!) results in a dramatic increase in the opacity. This results in very steep temperature gradients, leading to convection. In the Sun, for example, the outer 200,000 km (roughly one-third of its radius) is fully convective. In stars of even lower mass, the convective region can penetrate right down to the core.

The situation is exactly opposite in the upper part of the main sequence. In more massive stars, the outer regions are in radiative equilibrium. Their high surface temperature ensures that there are no steep temperature gradients in the outer layers. But their cores tend to be fully convective. This is because of steep temperature

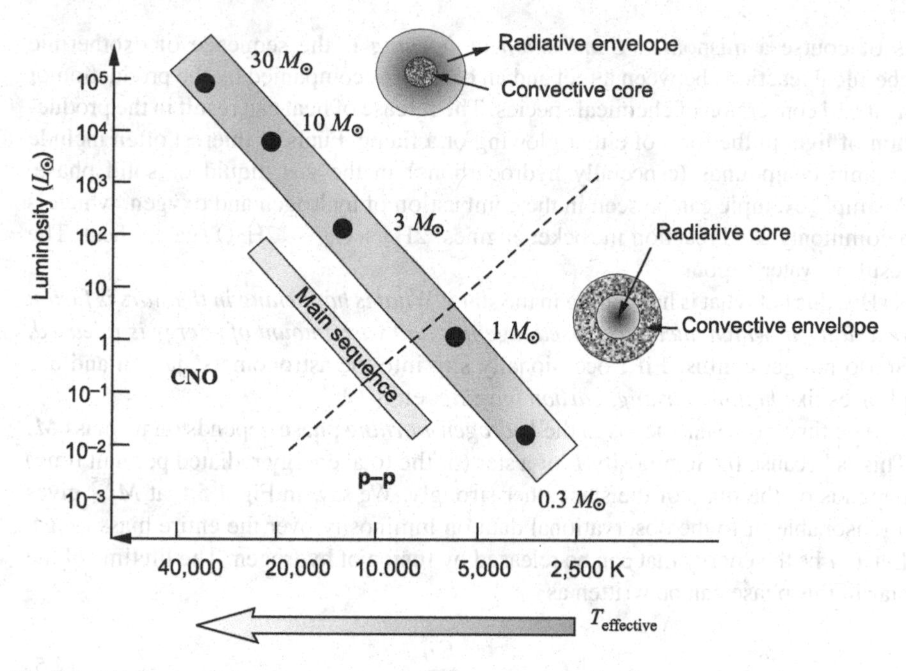

Fig. 2.6 This figure summarizes some important differences between the stars in the *upper part* of the main sequence and the *lower part*; the boundary is roughly around one solar mass

gradients in the core. The reason for this is the following. Remember that in these stars the CNO cycle is the dominant energy generation mechanism, and this process is very sensitive to temperature. Therefore, *the energy production more centrally concentrated, leading to steep temperature gradients.*

The above mentioned characteristics of stars in the main sequence are summarized in Fig. 2.6.

The Lifetime of Stars

As mentioned above, the stars on the main sequence are infant stars. The more massive among them would have formed only *recently* from interstellar gas. But some of these infants, like our Sun, were born a long time ago. We believe that that our solar system was formed roughly 4.5 billion years ago, but it is still in its infancy!

How long will the stars live? Perhaps it is more pertinent to ask how long the present act of the stellar drama will last. The main theme in this first act is the generation of energy by fusing hydrogen nuclei to form helium nuclei. This is often referred to as *hydrogen burning* in the astronomical literature.

[At this stage, let me make a parenthetical remark. After resisting for long (!), I have succumbed to the astronomers' terminology of *hydrogen burning*. This is of course a misnomer. *Combustion* or *burning* is the sequence of exothermic chemical reactions between a fuel and an oxidant accompanied by the production of heat and conversion of chemical species. The release of heat can result in the production of light in the form of either glowing or a flame. Fuels of interest often include organic compounds (especially hydrocarbons) in the gas, liquid or solid phase. A simple example can be seen in the combustion of hydrogen and oxygen, which is a commonly used reaction in rocket engines: $2H_2 + O_2 \rightarrow 2H_2O$ (gas) + heat. The result is water vapour.

But this not what is happening in the stars! *What is happening in the stars is fusion reactions, in which nuclei are fused together and vast amount of energy is released.* So do not get confused if I occasionally slip into the astronomers' jargon and use phrases like *helium burning, carbon burning*, etc.]

The time τ_H a star spends in the *hydrogen-burning* phase depends on its mass M. This is because the luminosity L of a star (or the total energy radiated per unit time) depends on the mass of the star rather strongly. We saw in Fig. 1.5 that $M^{3.5}$ gives a reasonable fit to the observational data on luminosity over the entire mass range. Let E_H be the energy that can be released by fusion of hydrogen. The lifetime of the star in this phase can be written as

$$\tau_H = \frac{E_H}{L}. \tag{2.5}$$

Let us assume for simplicity that the same fraction of the total mass of the star is consumed in this phase in all stars. We then have $E_H \propto Mc^2$ and

$$\boxed{\tau_H \sim \frac{M}{L} \sim M^{-2.5}.} \tag{2.6}$$

The Sun has already spent 4.5 billion years on the main sequence and it will be another 6.5 billion years before it begins the second act and leaves the main sequence. So *the hydrogen burning lifetime of the Sun is* $\sim 10^{10}$ *years*. The main sequence lifetime of stars of different mass can be estimated using this normalization and the scaling relation (2.6). Today, with fast computers at our disposal, one can actually calculate the lifetime by making some specific assumption about the abundance of hydrogen in the stars at the beginning of their lives. The main sequence lifetimes from such calculations is given in the table below.

$M/M_\odot$	1	4	5	6	7	8
$\tau_H/10^7$ years	700	8	4.9	3.3	2.5	2

The Ultimate Fate of the Stars

Once the hydrogen in the *core* is exhausted, the curtain will come down on the first act, and the subsequent acts of the stellar drama will follow. As we shall see, the subsequent acts will be of shorter and shorter duration. *Astonishingly, for a drama that has gone on for tens of millions or billions of years, the final act will last only a day or so!* Instead of continuing with the story of the life history of stars in a logical manner, we shall straightaway come to the point when the curtains come down at the end of the last act.

As I remarked while concluding the first chapter, this book is devoted to the question 'What is the ultimate fate of the stars?' What will happen when the nuclear reactions cease either because the central region is not hot enough or because it has run out of fuel. Since no more heat will be generated, the star will be in a serious predicament when the fossil heat is also radiated away.

Can the stars find peace?

It turns out that astronomers were confronted with this question way back in 1924, many decades before one had a satisfactory understanding of the evolution of stars. And they came up with some extraordinary answers.

Let us therefore go backwards in time to 1924 and get a historical perspective.

Chapter 3
White Dwarf Stars

The Strange Companion of Sirius

Eddington's theory of stars was a great success. To recall, this theory was predicated on the assumption that stars are globes of *ideal gas* in *radiative equilibrium.* The spectacular agreement between many of the predictions of this theory and observations lulled astronomers into thinking that that last word on the subject had been said. This feeling was shattered in 1924 when the American astronomer **Walter Adams** made a remarkable discovery regarding the companion of Sirius. But we are jumping the story! Let us go back and trace the history of this fascinating discovery.

Sirius is the brightest star in the night sky. You may be familiar with the very conspicuous constellation of *Orion* (the hunter). Sirius (the dog) is very close to this constellation. Since it is a very bright star, astronomers used it, along with other bright stars, to determine the time and set the clocks by. But astronomers noticed that Sirius was not a good *clock star* because its motion in the sky was a bit *jerky*. Let me explain. Stars are not stationary in the Galaxy; they have velocities. This causes their position in the plane of the sky to change. This is known as *proper motion.* Under normal circumstances, one would expect this proper motion to be linear in the sky. This was not the case with Sirius. In 1844, the great German astronomer and mathematician **Friedrich Wilhelm Bessel** deduced that Sirius was describing an elliptical orbit. Obviously there must be something for it to go around! Bessel thus deduced that Sirius must have an *invisible binary companion*, with each star moving around a common centre of mass. Bessel conjectured that the orbital period of the star must be about *half a century*. The modern value of the orbital period is 50.09 years! One should not be surprised by this since Bessel was an extraordinary astronomer and a great mathematician.

Eighteen years later, the *invisible companion* was actually seen by Alvan Clark. In 1862, Clark discovered a faint companion to Sirius while testing a new telescope. The brightness of the companion was 10^{-4} times smaller than the brightness of Sirius itself. Let us call Sirius and its companion Sirius A and Sirius B, respectively (Fig. 3.1). But the mass of Sirius B, deduced from the orbital period of Sirius A using

G. Srinivasan, *Life and Death of the Stars*, Undergraduate Lecture Notes in Physics,
DOI: 10.1007/978-3-642-45384-7_3, © Springer-Verlag Berlin Heidelberg 2014

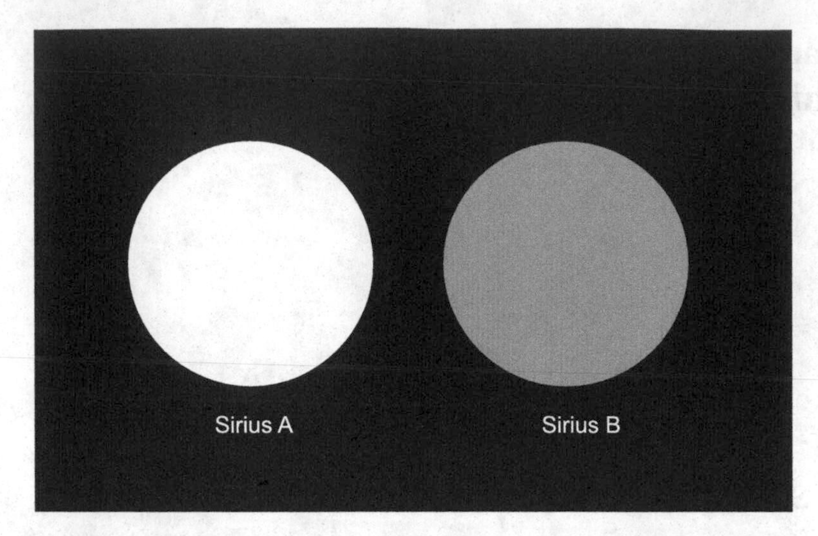

Fig. 3.1 The companion of Sirius turned out to be a very faint star, with brightness ten thousand times smaller than the brightness of Sirius. But the mass of the companion was similar to that of Sirius. This suggested that the companion must be a cool star, with a surface temperature much less than that of Sirius

Kepler's Law, was roughly the mass of the Sun. Since the mass of Sirius A was also roughly a solar mass, it was reasonable to suppose that the radii of Sirius and its companion must be roughly the same. The *faintness* of the companion, relative to Sirius A, could easily be accounted for if the surface temperature of the companion was *less* than that of Sirius A; in other words, if the companion was a *red star*, instead of being a *white* star like Sirius A. Remember that the *luminosity* or the total amount of energy radiated by an opaque body per unit time is given by

$$L = \text{surface area} \times \sigma T^4 = 4\pi R^2 \times \sigma T^4, \tag{3.1}$$

where R is the radius and T the temperature of the body. The constant σ is the *Stefan–Boltzmann* constant. Also recall that the *spectrum* of the radiation from an opaque body *peaks* at a wavelength which is determined by its temperature (Wien's Displacement Law). If you look carefully at the stars in the sky you will find that some are *bluish*, while some are *red.* This tells you immediately that the bluish star is *hotter* than the reddish star. Let us now return to the faint companion of Sirius. Since there is no reason to think that its radius would be very different from that of Sirius, one would expect the companion to be a red star; the surface temperature of the companion has to be much smaller to account for its low brightness (see Eq. 3.1). But there is a way to check this by measuring the colour of the star.

To determine the *colour* of a star, or equivalently its surface temperature, one has to measure the spectrum of the light from the star. This was done by Walter Adams in 1914 using the famous 100-inch telescope at the Mount Wilson Observatory in

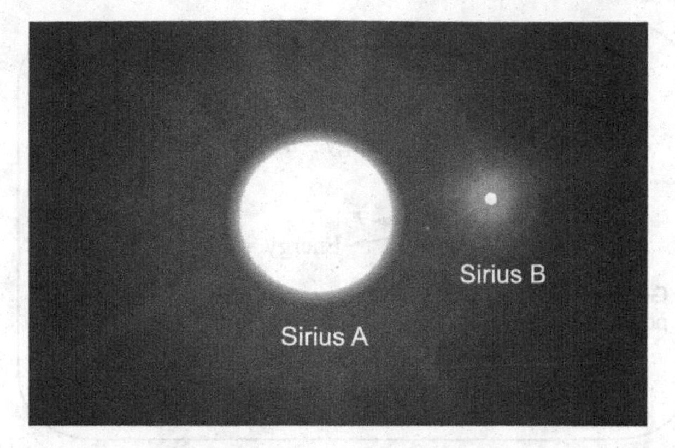

Fig. 3.2 Spectroscopic observations carried out by Walter Adams in 1914 showed that the surface temperature of the companion was roughly the same as that of Sirius, contrary to what was thought earlier. This implied that the radius of the companion of Sirius must be a hundred times smaller than that of Sirius. This, in turn, implied that *the mean density of the companion must be roughly a million grams per cubic centimetre*!

America (the world's largest telescope at that time). Surprise, surprise. The companion of Sirius was a *white hot star* (like Sirius) and *not a red star*! This spelt trouble. To see this let us go back to Eq. (3.1) and rewrite it as follows.

$$\frac{L_B}{L_A} = 10^{-4} = \frac{R_B^2}{R_A^2} \times \frac{T_B^4}{T_A^4}. \tag{3.2}$$

If the surface temperatures of the two stars are roughly the same, then the fact that Sirius B is 10^{-4} times fainter than Sirius A implies that its surface area must be roughly 10^{-4} times smaller. Equivalently, the *radius of* Sirius B *must be 100 times smaller than the radius of* Sirius A. In other words, Sirius B must be roughly the size of the Earth! This, however, would imply that the mean density of Sirius B must be 10^5–10^6 g cm^{-3} (Fig. 3.2). This appeared to be *nonsensical*, as Eddington put it.

Such an incredibly high density might defy comprehension, but the above conclusion was not nonsensical. This was established by Adams in 1924. In a very challenging observation, he tried to kill two birds with one stone. He set out to test one of the major predictions of Einstein's General Theory of Relativity and, at the same time, measure the radius of the companion of Sirius.

Gravitational Redshift

One of the important predictions of the General Theory of Relativity concerns the propagation of light in a gravitational field. In Newtonian theory of gravity only masses are affected by gravity. In Einstein's theory, all forms of energy contribute

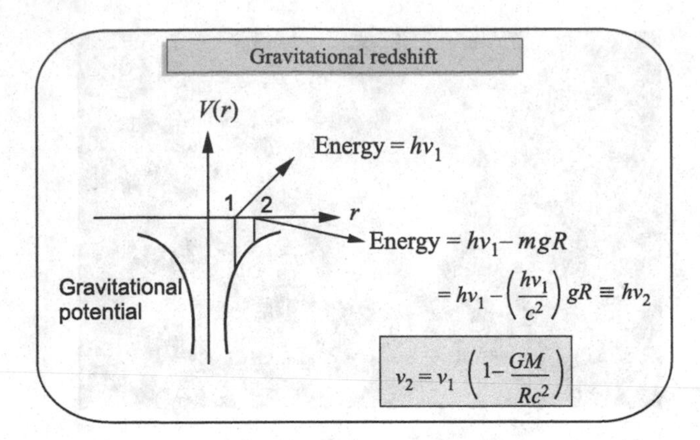

Fig. 3.3 One of the important predictions of Einstein's General Theory of Relativity is that gravitational field will affect the propagation of light. Just as the kinetic energy of a stone thrown *upward* from the Earth decreases with increasing height, and consequently *slows down*, similar thing must happen to light also. Einstein's theory predicts that *the frequency of radiation will decrease as it climbs out of a gravitational potential well.* A simple-minded argument, intended to make this plausible, is given in the figure

to gravity and are, therefore, affected by gravity. Since light is a form of energy, it is to be expected that light is also affected by gravity. Imagine a body of mass M and radius R. Let an atom on its surface emit radiation of wavelength λ or frequency ν. As the light propagates outwards, *its wavelength will be stretched.* Since the wavelength becomes longer one refers to it as a *redshift*. You will be familiar with a similar phenomenon when the source of light is moving away from the observer. In that case, one refers to it as Doppler shift. Here, since gravity is responsible for the red shift, one refers to it as *gravitational redshift.* Although this result is a prediction of the General Theory of Relativity, one can anticipate this result within the premise of the Special Theory of Relativity itself. Since it is difficult to deal with the wave nature of light in Newtonian gravity, we shall switch to the *photon* picture. You will recall that in 1905 Einstein introduced the revolutionary idea that light energy comes in *bundles* which have come to be known as *photons*. The energy of a photon is determined by the frequency of radiation and is given by the famous expression

$$E = h\nu, \tag{3.3}$$

where h is Planck's constant. Using the Special Relativistic expression, $E = mc^2$, one might call $(h\nu/c^2)$ as the *effective mass* of the photon. Now we are ready to attempt a poor man's derivation of the gravitational red shift.

When we throw up a stone, it slows down. As you know, this happens for the following reason. As the stone goes up, its potential energy increases. Consequently, its kinetic energy decreases. Since the kinetic energy is determined by the velocity, the stone slows down. A similar thing must happen to light. See Fig. 3.3.

Consider a photon climbing up the gravitational potential well from a radial distance 1 from the centre to 2. Let the initial energy of the photon be $h\nu_1$. When it reaches the radial position 2, its energy will be $h\nu_1 - mgR$, where the second term is the increase in the potential energy. Now let us substitute $(h\nu/c^2)$ for the *mass* of the photon, and $g = GM/R^2$ for the acceleration due to gravity. A little simplification shows that when the photon climbs from position 1 to 2, its frequency has decreased by the expression given below:

$$\nu_2 = \nu_1\left(1 - \frac{GM}{Rc^2}\right). \tag{3.4}$$

This is gravitational redshift. If we had cast the result in terms of the wavelength, we would find that the wavelength has been stretched. Note the essential difference between this and a stone climbing the potential well. The kinetic energy of the stone is determined by its velocity. Therefore, the velocity decreases as the kinetic energy decreases. Light cannot slow down! The energy of the photon is related to its frequency. As the energy of the photon decreases, *its frequency must decrease*.

The above result was obtained in a very heuristic manner. We used Newton's theory of gravity, the *corpuscular theory of light* due to Einstein and the Special Theory of Relativity. Since gravitational redshift is a manifestly General Relativistic effect, one must use the General Theory for a proper treatment. Well, this was done by the great German physicist and astronomer **Karl Schwarzschild** in 1915. The exact result for gravitational redshift is given below.

$$\boxed{\nu_\infty = \nu_0\left(1 - \frac{2GM}{Rc^2}\right)}$$

$$\boxed{\lambda_\infty = \frac{\lambda_0}{\left(1 - \frac{2GM}{Rc^2}\right)}} \tag{3.5}$$

Surprisingly, our simple-minded *derivation* of Eq. (3.4) differs from the exact result (3.5) only by a factor of 2 inside the parentheses. Notice two things in the above result.

1. The larger the mass, the larger is the redshift.
2. The smaller the radius, the larger is the redshift.

There is one other magical result implicit in Eq. (3.5). *When the radius of the object is precisely equal to* $(2GM/c^2)$, the *frequency of the photon goes to* zero *and the wavelength goes to* ***infinity***! We shall return to this most remarkable result in the next book of this series entitled, ***Neutron Stars and Black Holes.***

Experimental Verification of Gravitational Redshift

Let us now return to the historic observation by Adams in 1924 using the 100-inch telescope at Mount Wilson Observatory. His main objective was to test the important prediction by Einstein of gravitational redshift of spectral lines. Why did he choose the companion of Sirius? It is so faint that that any spectral line observations would have been incredibly difficult. Why not choose a nearby star like the Sun? A look at Eq. (3.5) will provide the answer. The *magnitude* of the red shift predicted by Einstein's theory depends upon the value of (M/R). While the mass of Sirius B was comparable to the solar mass, Adams had concluded from his 1914 observations that its radius might be a hundred times smaller than that of the Sun. Consequently, the gravitational redshift would be much larger. This is why he chose the companion of Sirius to test Einstein's prediction. The importance of verifying Einstein's prediction of gravitational redshift cannot be overstated. In a letter to Eddington written on 15 December 1919, Einstein said,

> If it were proved that this effect does not exist in nature, then the whole theory would have to be abandoned.

Using the estimated values for the mass and radius of Sirius B, Adams calculated the gravitational redshift that Einstein's theory would predict (see Eq. (3.5)). For his observation, he chose the Hβ and Hγ lines of the hydrogen atom. The name of the game was to accurately measure the wavelength of these lines in the spectrum of light from Sirius B and compare them with what quantum theory of atoms would predict. Since one knows the predicted wavelengths of these spectral lines very accurately, such a comparison would reveal whether there is a shift in the wavelength. Well, there was. What more, the measured redshift agreed very well with the prediction of the General Theory of Relativity! One cannot overemphasize the importance of this confirmation of one of the key predictions of Einstein's theory of gravity. And there was a bonus. If one accepts General Relativity, then this observation could be regarded as a measurement of the radius of Sirius B! This is why Eddington said, 'Professor Adams had killed two birds with one stone!'

To summarize this discussion, the measurement of the gravitational redshift proved beyond doubt that the companion of Sirius was, indeed, a stellar mass object, *but of planetary size*!

Before proceeding further, here is a little puzzle for you. Walter Adams was undoubtedly a very clever experimentalist. An accurate determination of the redshift of the spectral lines from Sirius B was a great technical achievement. But how can we be sure that this redshift is caused by gravity. It could just as well be due to Sirius B moving away from us. The redshift would then be due to good old Doppler effect! Think about it.

A Stellar Paradox: Have the Stars Enough Energy to Cool?

The discovery that the companion of Sirius is a *white star* had earlier led us to conclude that it must be a planetary size star. This, in turn, led us to the conclusion that the mean density of this star must be close to a *million grams per cubic centimetre.* Although this sounds nonsensical, the conclusion of the gravitational redshift measurement has sealed all escape routes. *We have no option but to accept that there are objects in the sky with mean densities of the order of* $10^6\, g\, cm^{-3.}$

This conclusion rattled Eddington. As usual, he was way ahead of the others in realizing the fatal difficulty posed by such ultra dense stars. This is what Eddington said:

> I do not see how a star which has once got into this compressed condition is ever going to get out of it ... Their high density is only possible because of the smashing of the atoms, which in turn depends upon the high temperature. It does not seem permissible to suppose that the matter can remain in this compressed state if the temperature falls ... When the supply of subatomic energy fails and there is nothing to maintain the high temperature, then on cooling down, the material will return to the normal density of terrestrial solids. The star must, therefore, expand, and in order to regain a density a thousandfold less the radius must expand tenfold. Energy would be required in order to force out the material against gravity. Where is this energy to come from? ... . the white dwarf can scarcely be supposed to have had sufficient foresight to make special provision for this remote demand. Thus the star may be in an awkward predicament—it will be losing heat continuously *but will not have enough energy to cool down.*
>
> From *Stars and Atoms*, 1927
> Sir Arthur Eddington

'*The star will need energy to cool*'. Put differently, imagine a body continuously losing heat but with insufficient energy to grow cold! What is one to make of this extraordinary statement? In less-cryptic terms, and in simple language, this is what the paradox amounts to.

Let E_V denote the negative electrostatic energy of a unit volume of the white dwarf material. At the high pressures inside a white dwarf, this energy per atom is essentially the sum of all the ionization energies required to strip the atom of all the electrons. And let E_K denote the kinetic energy per unit volume of completely ionized matter. If such matter were released of the pressure to which it is subject, then it could expand and resume the state of the ordinary unionized matter only if

$$E_K > E_V.$$

Is it guaranteed that this inequality will always be obeyed? An estimate of the electrostatic energy is given by

$$E_V = 1.32 \times 10^{11} Z^2 \rho^{\frac{4}{3}}. \tag{3.6}$$

We shall not pause to derive this. I request you to accept this result which can be found in standard books on Electricity and Magnetism. The kinetic energy per unit volume is given by:

$$E_K = \frac{3}{2}\frac{Nk_BT}{V} = \frac{3}{2}\frac{k_B}{\mu m_H}\rho T = 1.24 \times 10^8 \frac{\rho T}{\mu}. \tag{3.7}$$

This is easy to see. You will recall that the *average energy* of the particles is $\frac{3}{2}k_BT$. Multiplying this by the total number of particles N and dividing by the volume V we get the desired result. We can also write the above result in terms of the mass density ρ by multiplying and dividing the number density $n = N/V$ by the *average mass of the particles*. Since the gas consists of electrons and nuclei of different species, one introduces the notion of the *mean molecular weight* μ. The number density and mass density are related by the expression:

$$n = \frac{\rho}{\mu m_H}. \tag{3.8}$$

(See Chap. 3 of ***What Are the Stars?*** for a more detailed discussion of this.)

If this white dwarf material were released of the pressure to which it is subject, it can resume the state of ordinary normal atoms only if $E_K > E_V$. Using Eqs. (3.6) and (3.7), it is easy to show that $E_K > E_V$ only if

$$\rho < (0.94 \times 10^{-3} T/\mu Z^2)^3. \tag{3.9}$$

Clearly, the above inequality would be violated if the density is sufficiently high. In other words, at sufficiently high density the star will not have enough energy to expand and cool. This is what Eddington meant!

How was this stellar paradox resolved? Read on!

Chapter 4
The Principles of Statistical Mechanics

Before discussing Fowler's resolution of Eddington's paradox, let us digress to understand the developments that led Fowler to this seminal idea. It is essential that we have some understanding of the basic ideas of quantum statistical mechanics. This would also prepare us to discuss Chandrasekhar's theory of white dwarfs and the subsequent developments.

Classical Mechanics

Let us begin with a few words about *classical mechanics*. The subject of motion of bodies has been erected on the foundation laid by Newton. Newton's laws of motion enable us to analyse a variety of problems, such as the motion and collision of billiard balls, planetary motion, etc. What more, one can do this with great precision. We are now able to launch a rocket which travels for many years, covers a distance of many millions of kilometres and lands an instrument on one of the moons of Jupiter or Saturn! If you think about it a little, you will appreciate what an incredible achievement this is.

The key thing about classical mechanics is this. If you were dealing with an individual particle or body, one can describe its motion with arbitrary precision—there is no restriction on the precision to which we can determine its position and momentum.

Statistical Mechanics

During the nineteenth century, physicists turned their attention to the study of gases. One knew that the constituents of a gas were atoms and molecules. These atoms constantly collide with one another, change their energy and direction of motion.

G. Srinivasan, *Life and Death of the Stars*, Undergraduate Lecture Notes in Physics,
DOI: 10.1007/978-3-642-45384-7_4, © Springer-Verlag Berlin Heidelberg 2014

Physicists were interested in calculating gross properties of a gas, such as the *pressure* it exerts on the walls of the container, its *compressibility*, *specific heat*, etc. The basic assertion of the physicists was that these gross properties should be explainable in terms of the motion of its parts, namely, atoms and molecules. For simplicity, they assumed that the gas they were trying to describe was in *thermal equilibrium*. The laws of mechanics which apply to substances in thermal equilibrium are called *statistical mechanics*.

Let us pursue the description of a gas in thermal equilibrium. The first thing to appreciate is that constant collisions introduce a serious complication. *It is no longer possible to state that a particular atom* (let us say it is painted red) *has a particular velocity, say,* 5.123456789 m/s. By the time you could determine its velocity, it would have collided with some other atom and changed its speed and direction. Given this, the only meaningful question one could ask is the following: how many atoms are there with velocities in some *range*, say, between 5.123 and 5.124, and so on? Mathematically put, one could ask: *what is the fraction of molecules with velocities between* v and $v + dv$?

Maxwell's Velocity Distribution

The Scottish physicist **James Clerk Maxwell** solved this major puzzle in 1852. Consider a gas in a box, and let us first consider motion in one of the three dimensions. Maxwell discovered that the probability that the particles will have a velocity between v and $v + dv$ is given by

$$f(v)dv = \mathrm{C}e^{-\left(\frac{\text{kinetic energy}}{k_B T}\right)} dv = \mathrm{C}e^{-\frac{mv^2}{2kT}}\, dv. \tag{4.1}$$

The constant of proportionality can easily be determined by recognizing that the integral of the above probability distribution over all velocities from $-\infty$ to $+\infty$ must be *unity*. Carrying out this integral, we find that $C = \sqrt{m/2\pi k_B T}$ (if you like mathematics, try to integrate Eq. (4.1) over all velocities and verify this result). Since the motion of the particles in the three directions is independent, the probability distribution for the three-dimensional velocity $\vec{v}$ is just the product

$$f(v_x, v_y, v_z)dv_x dv_y dv_z \propto e^{-\frac{m(v_x^2+v_y^2+v_z^2)}{2kT}} dv_x dv_y dv_z.$$

$$f(v_x, v_y, v_z)dv_x dv_y dv_z = \left(\frac{m}{2\pi kT}\right)^{\frac{3}{2}} e^{-\frac{mv^2}{2kT}} dv_x dv_y dv_z. \tag{4.2}$$

It is better to remember this expression written in terms of the *momentum*. One of the reasons for this is that the expression written in terms of the momentum would be valid in special relativity also, that is, when the particles have speeds close to the speed of light. If written in terms of velocity, the expression would not be valid in relativity. Since velocity and momentum are proportional ($p = mv$), the probability distribution in terms of momentum will also have the same structure, namely, proportional to $e^{-K.E./k_B T}$. Written in full, it will read as follows:

$$f(p)d^3p = \frac{1}{(2\pi mkT)^{\frac{3}{2}}} e^{-\frac{p^2}{2mkT}} dp_x dp_y dp_z. \tag{4.3}$$

Equations (4.2) and (4.3) are *probability distributions* for velocities and momenta, respectively. Let the volume of the container be V and the total number of particles be N. The probability function tells us about the fraction of particles in a velocity or momentum range. A gas in thermal equilibrium will be uniformly distributed in the box. Therefore, the fraction of particles in a given velocity range will be the *same in every unit volume*; this is one of the requirements of the laws of thermodynamics. Therefore, the number of atoms or molecules per unit volume with velocities in the range dv_x, dv_y, dv_z, or momenta in the range dp_x, dp_y, dp_z, takes the form:

$$\boxed{N(v)dv_x dv_y dv_z = \frac{N}{V}\left(\frac{m}{2\pi kT}\right)^{\frac{3}{2}} e^{-\frac{mv^2}{2kT}} dv_x dv_y dv_z,} \tag{4.4}$$

$$\boxed{N(p)dp_x dp_y dp_z = \frac{N}{V}\frac{1}{(2\pi mkT)^{\frac{3}{2}}} e^{-\frac{p^2}{2mkT}} dp_x dp_y dp_z.} \tag{4.5}$$

The above two expressions have been *normalized* for N/V particles per unit volume. What this means is the following:

$$\iiint_{-\infty}^{\infty} N(v)dv_x dv_y dv_z = \frac{N}{V}\iiint_{-\infty}^{\infty}\left(\frac{m}{2\pi kT}\right)^{\frac{3}{2}} e^{-\frac{mv^2}{2kT}} dv_x dv_y dv_z = \frac{N}{V}. \tag{4.6}$$

The integral on the right-hand side is clearly equal to unity, since the probability distribution has been properly normalized. If you would like to convince yourself of this, go back to Eq. (4.1), and recall how the proportionality constant was derived.

The distribution described by Eq. (4.4) is the celebrated *Maxwell's velocity distribution*. The derivation of this was undoubtedly one of the great hallmarks of nineteenth century physics. Maxwell, of course, went on to make many more great discoveries. The greatest among these was the discovery of the equations describing the electric and magnetic fields—a discovery that puts Maxwell on the same pedestal as Einstein in the history of physics.

The Distribution of Speeds

Often we are not interested in the *direction* of the velocity vector, but only in its *magnitude*. The magnitude of the velocity vector is, of course, the speed. In Eq. (4.4) the probability involves only the square of the velocity and no change needs to be made. The only factor in Eq. (4.4) that contains information about the *direction* of the velocity is the infinitesimal volume $dv_x dv_y dv_z$ in *velocity space.* To derive Maxwell's distribution of *speeds* from the distribution of *velocities* all we have to do is rewrite the expression the volume in velocity space with velocity between $\vec{v}$ and $\vec{v}+d\vec{v}$ in a form in which the information about the direction of the velocity vector is erased. Well, all we have to do is to throw away the information about the direction of the velocity vector. This is easily done! You may have encountered the *spherical polar coordinate system* where the three coordinates replacing (x, y, z) are (r, θ, φ). Here, $r = \sqrt{x^2 + y^2 + z^2}$. In this coordinate system, the volume element $dxdydz$ becomes $r^2 dr \sin\theta d\theta d\varphi$. In our case,

$$dv_x dv_y dv_z = v^2 dv \sin\theta d\theta d\varphi. \tag{4.7}$$

Since we do not want the angular information (remember that the angles θ and φ contain the information on the direction), we shall get rid of it by allowing θ to have every possible value from $\pi/2$ to $-\pi/2$ and ϕ to have all values from 0 to 2π. This amounts to integrating $\sin\theta d\theta d\varphi$ over the range just mentioned. This will simply give us 4π. This should not surprise you since the solid angle of a sphere is 4π. If you are not comfortable with the above discussion, Fig. 4.1 should make this clear. Consider the shell between the concentric spheres with radii v and $v + dv$, respectively. Every vector having one end at the origin and of length in the range v and $v + dv$ will have its end point in this shell. And the volume of this shell is $4\pi v^2 dv$.

Therefore, the distribution of *speeds* (or the modulus of the momentum) is given by

$$\boxed{N(v)dv = \frac{N}{V}\left(\frac{m}{2\pi kT}\right)^{\frac{3}{2}} e^{-\frac{mv^2}{2kT}} 4\pi v^2 dv,} \tag{4.8}$$

$$\boxed{N(p)dp = \frac{N}{V}\frac{1}{(2\pi mkT)^{\frac{3}{2}}} e^{-\frac{p^2}{2mkT}} 4\pi p^2 dp.} \tag{4.9}$$

Please remember that in Eq. (4.8) the allowed values of the speed are from 0 to ∞. The magnitude of the momentum in Eq. (4.9) is over a similar range. Let us now concentrate on Eq. (4.9) written in terms of the momentum, and rewrite it by suppressing unnecessary details:

$$N(p)dp = \mathrm{C}e^{-\frac{p^2}{2mkT}} 4\pi p^2 dp. \tag{4.10}$$

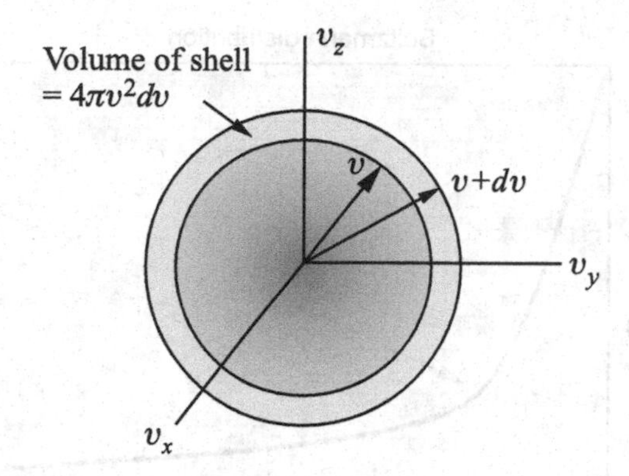

Fig. 4.1 In classical statistical mechanics, all values of velocity (or momentum) are allowed. Consider a certain value of the speed v. Since the velocity vector $\vec{v}$ can point in any direction, the number of allowed values of speed less than v is simply equal to the volume of the sphere with radius equal to v. Similarly, the number of allowed values of speed between $\vec{v}$ and $\vec{v} + d\vec{v}$ is equal to the volume of the shell $4\pi v^2 dv$. Note that this figure could also have been labelled with the components of the momentum

The right-hand side of Eq. (4.10) is a product of two factors. The first is the *probability* that the magnitude of the momentum will have a value in the range p and $p + dp$. Let us call this $f(p)$,

$$f(p) = Ce^{-\frac{p^2}{2mkT}}. \tag{4.11}$$

The second factor $(4\pi p^2 dp)$ in Eq. (4.10) is the volume of momentum space in the range p and $p + dp$. Let us call this the *density of states,* given by the expression $g(p)dp$. Using these two definitions, Eq. (4.10) can be rewritten as follows:

$$\boxed{N(p)dp = f(p)g(p)dp}, \tag{4.12}$$

$$\boxed{N(p)dp = \text{probability distribution} \times \text{density of states.}} \tag{4.13}$$

Written in this form, the expression is sufficiently general, so that it can be used for classical systems as well as quantum systems that we shall soon discuss. Also, it is very easy to remember!

Maxwell–Boltzmann Distribution

The fundamental discovery by Maxwell that we have just discussed was carried forward by the great Austrian physicist **Ludwig Boltzmann**. He laid the foundation and created the subject of statistical mechanics. Boltzmann recognized that the

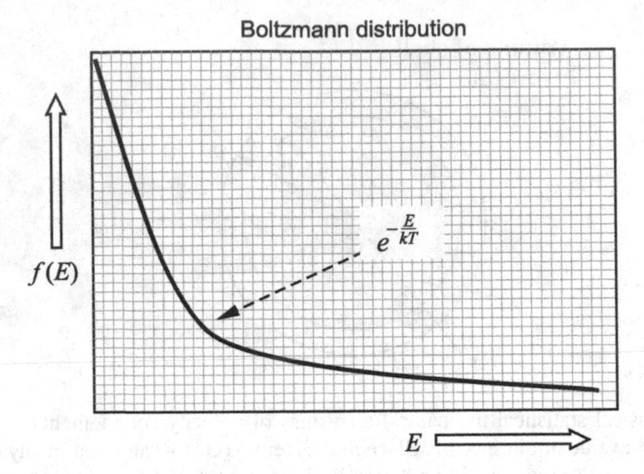

Fig. 4.2 In classical statistical mechanics, the probability $f(E)$ that a particle will have energy E is given by the Boltzmann distribution shown in the figure. It is an exponential distribution with a characteristic scale defined by $k_B T$, where the constant k_B is known as Boltzmann's constant

probability distribution for velocities discovered by Maxwell was far more general. In the problem we have been discussing, the atoms of the gas had only one degree of freedom, namely, the *translational degree of freedom.* The only form of energy that the atoms had was kinetic energy. If one was dealing a gas of molecules, then such molecules could have additional degrees of freedom, such as *rotational degree of freedom, vibrational degree of freedom*, and so on. There will be energy associated with each of these degrees of freedom. We now know that atoms have internal structure. The electrons in the atoms could be in various quantum levels with different energies.

Boltzmann was able to show that, in general, the *probability distribution of energy* of a system in thermal equilibrium is given by:

$$\boxed{f(E) \propto e^{-\frac{E}{k_B T}}.} \tag{4.14}$$

The constant k_B in the above expression is now known as *Boltzmann's constant.* Remember that for $f(E)$ to be a true probability distribution it should be *normalized* so that the integral of $f(E)$ over all energies is unity. See Fig. 4.2.

Let us next write the expression for $N(E)dE$, the average number of particles with energies between E and $E + dE$. This can be obtained from Eq. (4.10). In Newtonian mechanics, energy and momentum are related by $E = p^2/2m$. All one has to do is to express the density of states in terms of the energy. It may be verified that

$$p^2 dp = \sqrt{2m^3}\sqrt{E}dE. \tag{4.15}$$

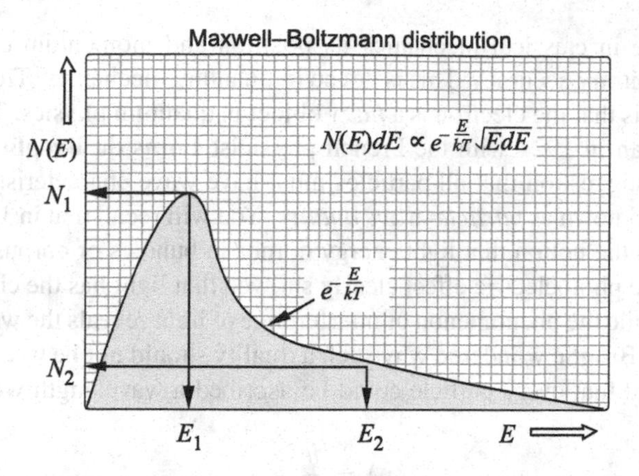

Fig. 4.3 The figure shows the number of particles with energy between E and $E+dE$ in Boltzmann statistics, as shown in Eq. (4.16). This is referred to as the Maxwell–Boltzmann distribution, or simply the Boltzmann distribution. The number distribution is a product of the probability $f(E)$ that a particle will have energy E and the density of states $g(E)dE$, which is the number of energy values between E and $E+dE$. As may be seen from Eq. (4.15), $g(E)dE \propto \sqrt{E}dE$. For $E \ll kT$ the above distribution is proportional to $\sqrt{E}$, while for $E \gg kT$ it decreases exponentially. This exponential tail is characteristic of classical statistics

Therefore, Eq. (4.10) written in terms of the energy would read as given below:

$$\boxed{N(E)dE = (\ldots)e^{-\frac{E}{kT}}\sqrt{E}dE.}$$

Restoring all the constants, we get:

$$\boxed{N(E)dE = \frac{N}{V}\left(\frac{2}{\sqrt{\pi(kT)^3}}\right)e^{-\frac{E}{kT}}\sqrt{E}dE.} \tag{4.16}$$

Figure 4.3 shows a sketch of this important distribution. As an exercise, try to verify that for small energies this function increases as $\sqrt{E}$, while it decreases exponentially at large energies as $e^{-E/kT}$.

Quantum Mechanics

Particles are Waves!

Before discussing the modification of Boltzmann's statistical mechanics in quantum theory, it would be useful to recall some of the salient *differences* between classical and quantum mechanics. As we mentioned before, classical mechanics is deterministic. That is to say that the motion of an electron, for instance, can be described precisely, since in classical mechanics its position and momentum can be determined with arbitrary accuracy. This is not so in quantum mechanics. The underlying reason for this is that the electron is a *fuzzy* object in quantum physics. The quantum revolution began in 1924 with the French physicist **Louis de Broglie** making the extraordinary suggestion that all particles must have wave characteristic also. This has come to be known as *particle–wave duality*. You will recall that in 1905 Einstein had introduced the notion that light energy comes in bundles or corpuscles, known as *photons*. The photoelectric effect clearly showed that light has the characteristics of particles, while the phenomenon of interference of light reveals the wave nature of light. Louis de Broglie wondered why such a duality should not be true for *particles* also! He argued that every particle could be ascribed a wavelength which is given by:

$$\lambda = \frac{h}{p}, \tag{4.17}$$

where p is the momentum of the particle. This is now called *de Broglie wavelength.* Notice that this equation holds for photons, as well as material particles; for the momentum of the photon is $h\nu/c = h/\lambda$.

The Wave Function

Louis de Broglie's idea is one of the principles on which the superstructure of quantum mechanics is built. If particles are waves then they must be solutions to some *wave equation.* This equation was discovered by **Erwin Schrödinger** and is named after him. Let us consider electrons to be specific. According to the wave mechanics developed by Schrödinger, each state of the electron (known as quantum states) represents a system of *standing waves*, or a *normal mode* of a harmonic vibration. This is exactly like the vibrations of a plucked string in its fundamental mode or its overtones.

Because an electron is a wave—and not a point particle—in quantum mechanics we can only talk of an electron as a *wave packet*. And where is the electron in relation to the wave packet? At any given instant, an observation might find the particle at any point where the wave function Ψ is different from zero. One is allowed to talk only in terms of probability. The probability of finding the particle in the neighbourhood of a point is given by $|\Psi|^2$, the square of the modulus of the wave function.

Heisenberg's Uncertainty Principle

The fuzziness of the particles in quantum mechanics destroys the determinism which is the hall mark of classical mechanics. This inherent indeterminacy in quantum physics was stated in a mathematical form by **Werner Heisenberg** in 1927. Let us say that we design an experiment to very precisely measure the *position* and *momentum* of an electron in a box. According to Heisenberg, one cannot do this. One is forbidden from measuring both the position *and* momentum with infinite precision. If you try to measure the position very accurately, then you will end up having a huge error in your measurement of the momentum. Similarly, you cannot measure the momentum very accurately without seriously compromising on the accuracy with which you can determine the position of the electron. Stated more precisely, if Δx is the error in the measurement of the position and Δp the error in the measurement of the momentum, then:

$$\Delta p \Delta x \geq \frac{h}{2}, \tag{4.18}$$

where h is Planck's constant. The wave–particle duality principle of de Broglie and Heisenberg's uncertainty principle are the two underlying axioms of quantum mechanics, as we know it today. Einstein never liked the uncertainty principle. He refused to accept it despite the incredible success of quantum mechanics during the three decades before his death in 1955. But that is a different story. If you would like to read more about the development of quantum mechanics I would like to strongly recommend to you the three volumes of *The Quantum Revolution*, by G. Venkataraman.

Discrete Energy Levels

A fundamental consequence of the wave nature of matter is that energy levels are discrete in quantum mechanics. You already know from Bohr's theory that the allowed energy values of an electron in an atom are discrete. We call them energy levels. The discreteness of the energy levels is quite general. Since this will be useful soon, let us consider an electron in a one-dimensional box of length, L, and impenetrable walls. In other words, the electron is confined to $0 < x < L$. The electron can move freely in this range but rebounds at the two walls. This is equivalent to the boundary condition that the wave function $\Psi = 0$ at $x = 0$ and at $x = L$. We have already mentioned that according to Schrödinger, each state of the electron corresponds to a *standing wave solution*. You know from the familiar example of the vibrations of a string held at two points that the normal modes correspond to sine waves with wavelength $\lambda = \frac{2L}{n}$, where $n = 1, 2, 3, \ldots$. The wave function of the first three states are shown in Fig. 4.4.

The allowed values of the energy, E, corresponding to these standing wave solutions are given by:

$$\boxed{E_n = \frac{n^2 h^2}{8mL^2}.} \tag{4.19}$$

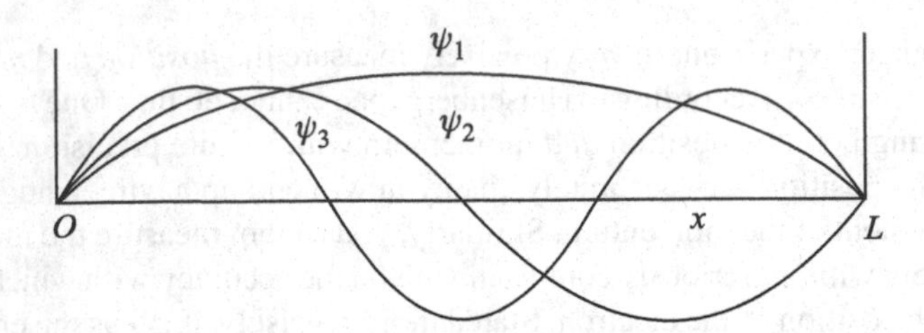

Fig. 4.4 The wave functions of the first three discrete energy levels of a particle in a one-dimensional box of length L. Notice that the wave functions satisfy the condition for standing waves. These are precisely the same as the normal modes of vibration of a string fixed at the two ends

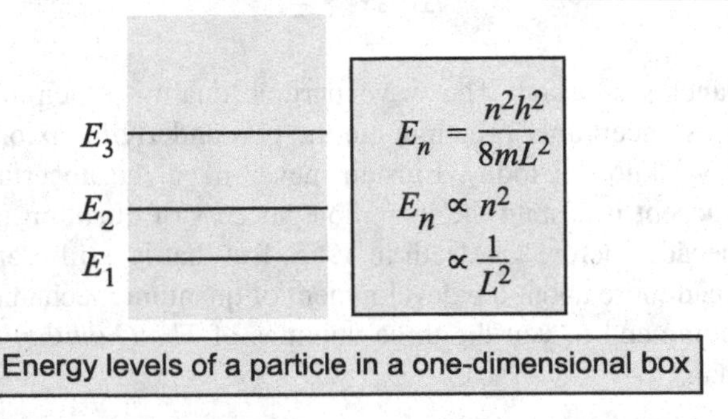

Fig. 4.5 The allowed values of the various discrete energy states are labelled by a quantum number, n, which can assume values $n = 1, 2, 3, \ldots$. Notice that the energy values are directly proportional to the square of this quantum number, unlike in the case of the hydrogen atom where it is inversely proportional to n^2. Notice also that the energy levels are determined by the size of the box

We have not derived this expression, but it can be found in all elementary texts on quantum mechanics (see Fig. 4.5). Please notice two features of the allowed energy values:

1. The allowed energy levels are discrete since $n = 1, 2, 3$, etc.
2. The smaller the box, the larger are the allowed values of the energy levels.

The above discussion can easily be generalized to electron in a three-dimensional box.

Quantum Statistical Mechanics

We are now ready to discuss the basic principles of *quantum statistical mechanics.* You will recall that the need for statistical mechanics first arose in the context of the kinetic theory of gases. To fix our ideas, let us mention two examples of problems where the need arises for statistical mechanics while dealing with quantum systems.

1. First, let us consider a gas consisting of hydrogen atoms (see Fig. 4.6). In classical physics, the atoms were point particles. We now know that the hydrogen atom has

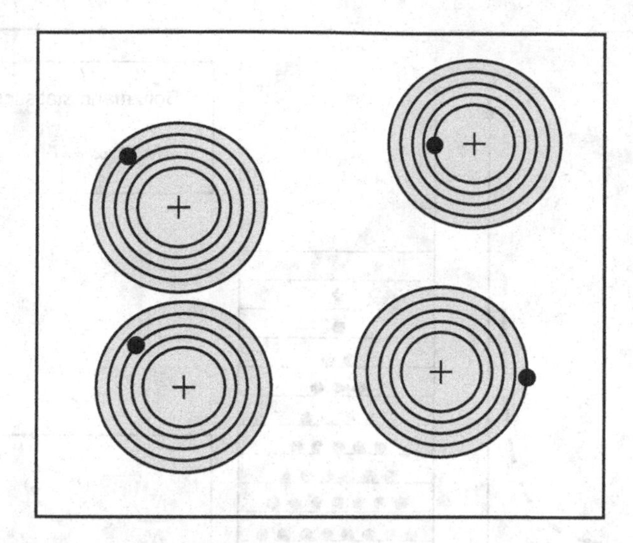

Fig. 4.6 An example to illustrate how statistical considerations enter quantum mechanics. Consider a gas of, say, hydrogen atoms at a temperature, T. When the gas is very dilute, all the atoms will be in their ground state, that is, the lowest energy state with the quantum number, $n = 1$. But if the density is sufficiently large, the atoms will frequently collide with one another. In the process, the kinetic energy of the atoms ($\sim\frac{3}{2}kT$) can be converted to internal energy of the atoms. As a consequence, all the atoms will no longer be in their lowest energy state, as illustrated in the figure. The inverse of this process can, and will, also happen during collisions. After sufficient number of collisions, excitation and de-excitation of the atoms, one can ask the following question: what is the average number of atoms with the electron in a particular energy level? Statistical mechanics is intended to answer questions such as this

an internal structure with an electron orbiting a proton at the centre. According to Bohr's theory, the orbiting electron can be in any of the allowed set of energy levels, defined by a quantum number n. In an *isolated* atom, the electron will generally be in the lowest quantum state with $n = 1$; the binding energy of this level is -13.6 eV. But in a gas, not all the atoms will be in this lowest level. At a finite temperature, the atoms in the gas will be constantly colliding with each other. The energy gained in one such collision can be used to *excite* the electron in that atom to a higher level. Whether it is excited to the level with $n = 17, 101$, or 272 will depend upon how much energy is gained in the collision. Collisions can also extract energy from an atom, causing an electron in, say, the $n = 272$ level to jump down to one of the lower levels. Now let us pose a question that arises in practical situations.

Let us consider a gas of N hydrogen atoms. We would like to know what fraction of the atoms are in a particular electronic energy level. In other words, we would like to know the average value of $N(E_n)$, where E_n are the internal energy levels of the atom and should not be confused with the kinetic energy of the atoms.

2. As a second example, let us consider a gas of N electrons confined to a volume V. As we discussed above, the energy of each electron is quantized, and can assume

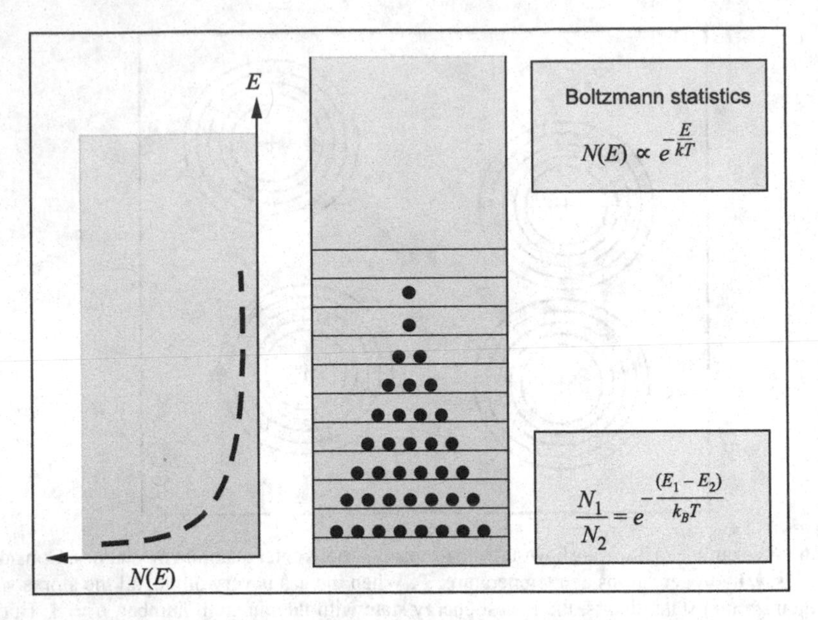

Fig. 4.7 A schematic representation of how energy levels are populated in Boltzmann statistics. In thermal equilibrium, the ratio of population in the various levels is uniquely determined. Indeed, this is how the concept of excitation temperature is defined. In true thermodynamic equilibrium, the temperature so defined will be the same as the kinetic temperature that enters Maxwell's velocity distribution

only a set of discrete values. Let us designate these discrete energy levels by E_n. Let the electron gas be in thermal equilibrium at a temperature T. Clearly, not all the electrons will have the same energy; one can safely anticipate that there will be a distribution of energies. The question is, 'What is the average number of electrons with a particular energy?'

For a sufficiently dilute gas, at a sufficiently high temperature, the answer to this question in both the examples mentioned above is given by Boltzmann's statistics (see Fig. 4.7).

Boltzmann Statistics and Quantum Systems

Let us assume that although we are dealing with energy levels which are quantized, the *distribution* of the given number of particles amongst these levels—the *statistics*, if you like—is still governed by *classical statistical mechanics*; in other words, the Boltzmann statistics that we discussed earlier. The average number of particles with energy equal to E_n is given by:

$$N(E_n) \propto e^{-E_n/kT}\,. \tag{4.20}$$

It follows that the ratio of the population in two energy levels is given by:

$$\frac{N_1}{N_2} = e^{-\frac{(E_1 - E_2)}{kT}}. \tag{4.21}$$

The Meaning of Temperature

There is another important way to look at this result. For a system in true thermal equilibrium, the ratio of populations given by Eq. (4.21) *defines* what we mean by *temperature.* Imagine that we are dealing with an assembly of atoms. Have you wondered what we mean by the gas being at a certain temperature? You would have first encountered the concept of temperature in thermodynamics. If you go back to your books you will find that the nineteenth-century physicists had not done a good job in defining the concept of temperature. This was set right once and for all in 1909 by the French mathematician Carathéodory. His formulation of thermodynamics was for the first time logically consistent. How is temperature defined in statistical mechanics?

Maxwell would have defined *temperature* as the *width of his velocity distribution.* For convenience we have reproduced below Maxwell's distribution of velocities, given by Eq. (4.4).

$$N(v)dv_x dv_y dv_z = \frac{N}{V}\left(\frac{m}{2\pi kT}\right)^{\frac{3}{2}} e^{-\frac{mv^2}{2kT}} dv_x dv_y dv_z$$

The function $e^{-mv^2/2kT}$ is a very well known function in mathematics and is called a Gaussian. It is a bell-shaped curve whose characteristic *width* is given by $\sqrt{kT/m}$. So Maxwell would have said that temperature is nothing but the width of the velocity distribution. Equivalently, he might have said that temperature is a measure of the average energy of the particles, which is given by $\frac{3}{2}kT$.

But Boltzmann would have disagreed. He would have said that the concept of temperature is defined by the ratio of population in different energy levels, defined by Eq. (4.21).

Who is right? Well, both of them are right, if our gas of atoms is in true thermodynamic equilibrium. In true thermodynamic equilibrium, frequent collisions between the particles will ensure that the different degrees of freedom *talk to each other,* as it were. Therefore, in true thermal equilibrium, the temperature derived by Maxwell and Boltzmann would be the same!

Are there situations where the two *temperatures* would *not* be the same? Yes. If the gas is extremely tenuous then there would not be sufficient number of collisions to establish true thermal equilibrium. And such situations are very common in astronomy. For example, the *interstellar medium* is very tenuous, with a number density of roughly 1 atom per cm^3! Compare this with a number density of roughly 10^{23} atoms per cm^3 in terrestrial matter! Not surprisingly, in such a tenuous gas:

- Matter and radiation will not come to thermal equilibrium. Consequently, if one were to ascribe a certain *temperature* to the radiation (let us call it the *radiation temperature*), it will *not* correspond to the temperature of the gas. In such situations, the spectrum of radiation will not correspond to that of a black body.
- The kinetic degree of freedom will not come to equilibrium with the internal degrees of freedom, such as the internal electronic levels, the vibrational levels, the rotational levels, etc. Consequently, the *kinetic temperature*(defined in terms of the width of the velocity distribution) will not be equal to the *excitation temperature* (defined in terms of the ratio of population among the internal levels).

A rose by any other name would still be a rose. But this is not so for temperature, except under conditions of true thermodynamic equilibrium!

Quantum Statistics

The rules of quantum statistics differ from those of classical statistics in three essential ways. We shall now discuss them.

Cells in Phase Space

A useful concept in statistical physics is that of *phase space*. This is a six-dimensional space, with three dimensions representing spatial co-ordinates (x, y, z), and the other three dimensions representing the three components of the momentum (p_x, p_y, p_z). Let the gas be contained in a volume V, and let us focus on the allowed values of the three components of the momenta.

In classical statistical mechanics, all values of the momenta are accessible to the particles; there was no restriction. For example, the number of momentum values in the interval p and $p + dp$ is $4\pi p^2 dp$. (Refer to Fig. 4.1. Although this figure has been labelled with the components of the velocity, it could also have been labelled with the momenta.)

In quantum statistical mechanics, not all values of the momenta are allowed—*only certain discrete values are allowed*. This is a simple consequence of the wave nature of matter and we saw that in our discussion of an electron in a box (see Figs. 4.4 and 4.5). Consequently, *momentum space* is not continuous but discrete. As may be seen from Fig. 4.8, in quantum physics *momentum space* is constructed using basic building blocks, or *cells*, whose volume is given by the expression:

$$\text{Volume of elementary cell} = \left(\frac{h}{L}\right)^3 = \frac{h^3}{V}, \tag{4.22}$$

where $V = L^3$ is the volume of our cubical box of length L.

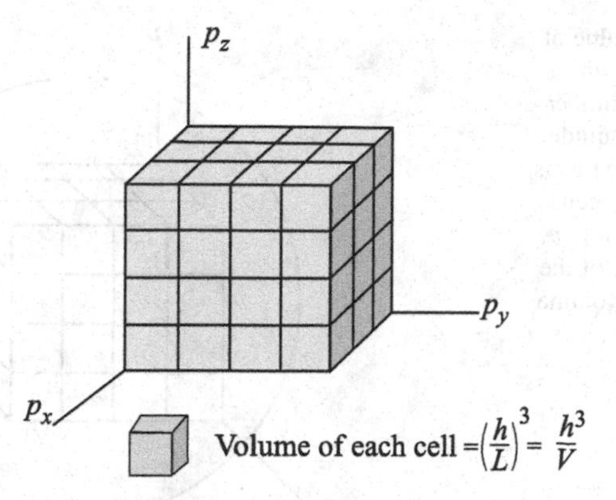

Fig. 4.8 In quantum mechanics only discrete values of momentum are allowed. We saw that in the example of a particle in a one-dimensional box, but this is true in general. Consequently, momentum space is not continuous as it is in classical physics (Fig. 4.1). Instead, it is made up by stacking cells. The volume of these primitive cells is determined essentially by Heisenberg's uncertainty principle. Larger the volume of the box, smaller is the uncertainty in the momentum, and therefore the cells are smaller. The bottom line is that *there is one momentum state per cell in phase space*

There is one allowed value of the combination (p_x, p_y, p_z) inside each of these cells. Let us say we are interested in the number of allowed momentum values less than a certain value p. In classical physics, this is simply the volume of the sphere of radius p, namely, $\left(\frac{4\pi}{3}p^3\right)$ *per unit volume of the box.* For a box of volume V, the number of allowed momentum values are $V\left(\frac{4\pi}{3}p^3\right)$. In quantum statistics, this number is equal to the number of cells within the sphere, namely,

$$\frac{\frac{4\pi}{3}p^3}{\left(\frac{h^3}{V}\right)} = \frac{V\frac{4\pi}{3}p^3}{h^3}. \qquad (4.23)$$

In Eqs. (4.12) and (4.13) we had introduced the notion of the *density of momentum states:*

$$g(p)dp = 4\pi p^2 dp.$$

This is the number of momentum values between p and $p+dp$. In quantum statistics, the density of states is given by:

$$\boxed{g(p)dp = \frac{V 4\pi p^2 dp}{h^3}.} \qquad (4.24)$$

The graininess of momentum space can also be understood in terms of Heisenberg's uncertainty principle. Let us consider the motion of the particle in a one-dimensional

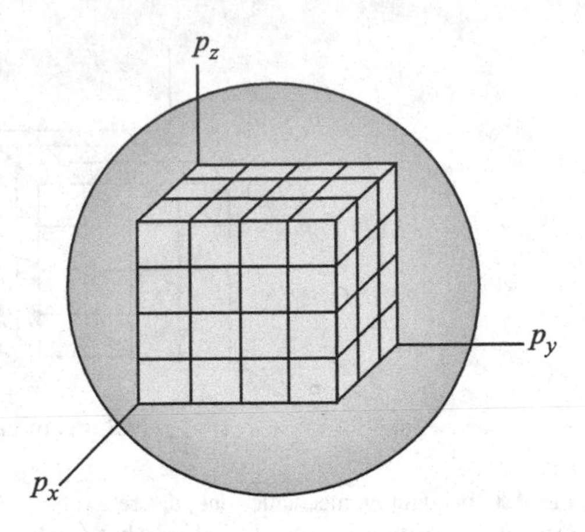

Fig. 4.9 Consider a value of the momentum vector whose magnitude is p. The number of states with the magnitude of momentum less than this is equal to the number of cells inside the sphere of radius p. This is just the volume of the sphere divided by the volume of the elementary cells, see Eq. (4.24)

box of length L. Recall that for the pair of variables *position* x and *momentum*, p, according to the uncertainty principle:

$$\Delta x \Delta p_x \geq h.$$

Since we *know* that the particle is inside the box of length L, there is an uncertainty in the momentum:

$$\Delta p_x \sim \frac{h}{L}.$$

In other words, we are not allowed to define the x-component of the momentum of the particle with accuracy better than h/L; *it is meaningless to do so*! The same argument applies to p_y and p_z. Therefore, the degree of graininess, if you like, is given by:

$$\Delta p_x \Delta p_y \Delta p_z \sim \left(\frac{h}{L}\right)^3.$$

We thus recover our earlier result that *there is only one momentum state per cell in phase space.* If one makes the box bigger, then the uncertainty in the momentum becomes smaller. The volume of the cell decreases, and the number of cells within a momentum range increases (refer to Figs. 4.8 and 4.9).

Indistinguishable Particles

Let us next discuss the second fundamental difference between classical and quantum statistical mechanics. Boltzmann derived his statistical distribution by assuming that the particles are *distinguishable.* The atoms or molecules of the gas have *identity cards,* even though they may be atoms of the same species. Boltzmann did not

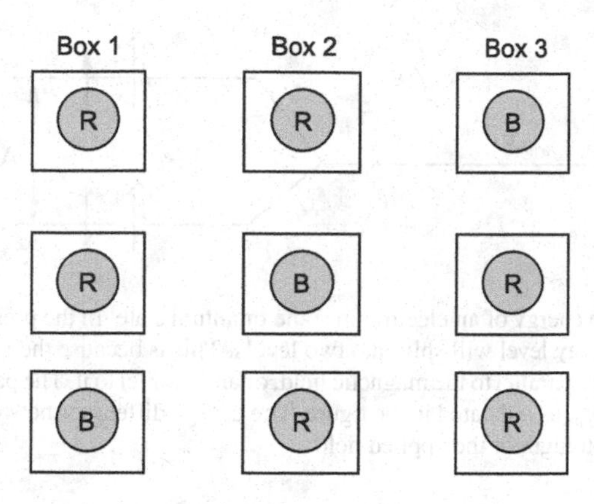

Fig. 4.10 This figure shows the number of ways in which *two identical red balls* and *one blue ball* can be distributed among three boxes, with one ball per box. If the *two red balls* are distinguishable (that is, not identical), then they could be labelled R_1 and R_2. Clearly, in each row of the above figure, there would be another configuration with R_1 and R_2 interchanged. Thus, there will be three more arrangements possible

know about the constituents of the atoms—electrons, protons and neutrons. But he would have assumed that they are distinguishable. In quantum statistics, elementary particles of a given species are *indistinguishable*. In other words, whereas you can distinguish between an electron and a proton, all electrons are to be regarded as *identical*. Similarly, all neutrons are identical.

Where does this distinction between distinguishable identical particles enter the discussion? Remember that the basic objective of statistical mechanics is to calculate certain probability distributions. For example, *we would like to know what the average number of particles with a certain energy is*. The fact that we are seeking an *average* implies that if we do repeated measurements we would get different answers. Let us say the total number of particles is N. The name of the game is to enumerate the number of possible ways of distributing these particles among M levels, and have a prescription to find the average occupancy in a given level. It is in this enumeration that the distinction between distinguishable and identical particles comes in. Let us consider a simple example. We want to distribute two *red balls* and one *blue ball* into three boxes, with one ball per box. The various possibilities are shown in Fig. 4.10 for *identical red balls*.

If the red balls are *distinguishable* (let us call them R_1 and R_2) then there would be three more arrangements; one more arrangement for each row, with R_1 and R_2 interchanged. This seemingly innocuous distinction between distinguishable and indistinguishable particles leads to very different statistics in quantum physics.

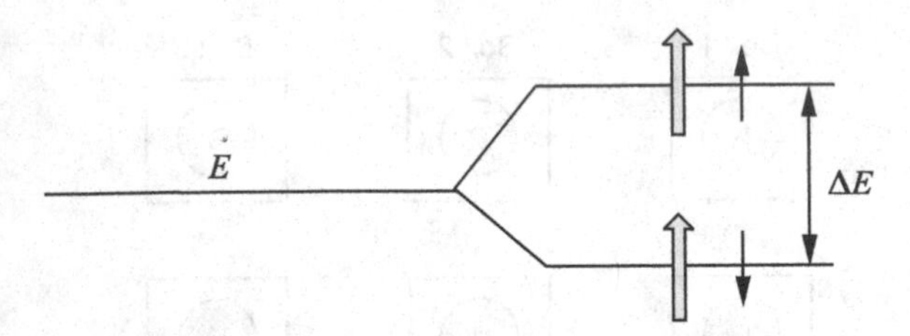

Fig. 4.11 Let E be the energy of an electron in some quantum state. In the presence of an external magnetic field, this energy level will split into two levels. This is because the magnetic moment of the electron can be either parallel to the magnetic field, or antiparallel to it. The parallel configuration will have a higher energy, as indicated in the figure. The energy difference between these two levels will depend upon the strength of the applied field

Spin and Statistics

The above discussion should suffice to appreciate the essential differences between classical and quantum statistical mechanics. In classical physics we had *Boltzmann statistics*. Similarly, is there a unique statistics in quantum physics? The answer is no. And that is because there are *two families* of particles in quantum physics. The rule of distributing N elementary particles among M levels or states of a quantum system depends upon which of the two families the elementary particles belong to. Elementary particles are grouped into these two families depending upon an important internal property of the particles known as *spin*. Let us discuss this a little.

In classical physics, an electron is characterized by its mass and charge. It can also have angular momentum by virtue of its motion around the nucleus in an atom, or due to its gyration in a magnetic field. We discussed this in the context of the *Zeeman Effect* in ***What Are the Stars?***. In 1925, Uhlenbeck and Goudsmit pointed out that certain features in atomic spectra could be explained if the electron possessed *intrinsic angular momentum* and *magnetic moment*. This angular momentum was not by virtue of any orbital motion; it was intrinsic to the particle. They initially thought that this might be due to the electron *spinning* or *rotating* about an axis through its centre of mass, like the rotation of a *top*. According to them, an electron is like a little *bar magnet*. Its magnetic moment could be oriented either parallel or antiparallel to an applied magnetic field. Obviously, the two orientations would have slightly different energy (see Fig. 4.11). Recall that two bar magnets in close proximity would have different energy depending on whether the magnetic fields of the two magnets are parallel or antiparallel. From a detailed analysis of atomic spectra, Uhlenbeck and Goudsmit deduced that the internal angular momentum of the electron is equal to $\frac{1}{2}\hbar$, where $\hbar = \frac{h}{2\pi}$. You may remember that Planck's constant has the dimensions of *angular momentum* (if you did not know it already, convince yourself of this).

The notion that this intrinsic angular momentum of the electron may be due to the electron *spinning* about its axis has serious difficulties. If this were true then the electron could have had any arbitrary spin angular momentum, depending upon how fast it is spinning. But atomic spectra clearly showed that this was not the

Table 4.1 Quantum numbers carried by some elementary particles

Particle	Spin	Fermion	Boson
Electron	1/2	√	
Positron	1/2	√	
Neutrino	1/2	√	
Proton	1/2	√	
Neutron	1/2	√	
μ meson	1/2	√	
Omega	3/2	√	
π meson	0		√
K meson	0		√
Photon	1		√
Graviton	2		√

case. The intrinsic angular momentum of the electron was equal to $\frac{1}{2}\hbar$, and not any arbitrary multiple of it. This unsatisfactory state of affairs was resolved in 1928 when the great English physicist **P. A. M. Dirac** discovered an equation which brought the wave-mechanical theory of the electron into harmony with Einstein's special theory of relativity. The earlier wave equation, discovered by Schrödinger, was not consistent with Special Relativity. The relativistic wave equation is now known as *Dirac Equation*, and its discovery by Dirac is regarded as one of the greatest achievements of twentieth century physics. The relevant point for us is that in Dirac's theory an electron is naturally endowed with an intrinsic angular momentum equal to $\frac{1}{2}\hbar$, and an associated magnetic moment.

Soon it became clear that *all* elementary particles must have this attribute of intrinsic angular momentum, referred to as *spin*. One thus ascribes a *spin quantum number, s*, to each species of particle. The associated spin angular momentum is $s\hbar$. This spin quantum number can take the values given below:

$$s = 0, \frac{1}{2}, 1, \frac{3}{2}, 2, \frac{5}{2}, \ldots \tag{4.25}$$

This might sound odd to you because the quantum number you would be familiar with—the *principal quantum number*, *n*, which Bohr introduced in the context of the hydrogen atom—assumed only *integral values* $n = 1, 2, 3, \ldots$. But that is the way it is. *The spin quantum number can be either an integer* $(1, 2, 3, \ldots)$ *or half-integer* $(\frac{1}{2}, \frac{3}{2}, \ldots)$.

Table 4.1 gives the spin quantum numbers of some of the more important elementary particles. Particles with half-integral spin quantum numbers are called *fermions* and particles with integral spin quantum numbers are called *bosons*.

Let us now get back to our story of quantum statistical mechanics. We said that the rule of populating *N* elementary particles among *M* levels or states depends upon the family to which the particles belong. Fermions and bosons are the two families of particles that we had in mind. If the particles under consideration are *fermions* (like electrons, protons and neutrons) then the rule of populating them is governed by what is known as *Fermi–Dirac Statistics*. On the other hand, if the particles are *bosons* then they obey a different quantum statistics, known as *Bose–Einstein Statistics*. In this volume we shall be mainly concerned with electrons, protons and neutrons. Therefore we shall devote the whole of Chap. 5, 'Fermi–Dirac Distribution', to a discussion of how Fermi–Dirac statistics can be applied to fermions.

Before we move on, a few words about Bose–Einstein statistics would be in order. This statistics was discovered by **S. N. Bose** working in Calcutta. His main objective was to derive the frequency spectrum of black body radiation, now known as Planck's Law, from fundamental principles. Although Planck had discovered the nature of the spectrum, he had not provided a sufficiently satisfactory derivation. Bose approached the problem of deriving Planck's law as a problem of discovering the statistical distribution of photons in thermal equilibrium with matter. He was able to do this in 1924. His fundamental results were contained in two papers, which he sent to Einstein, requesting him to translate them into German, and get them published in a prestigious German Journal. Einstein did that! But even before he received the second paper, Einstein realized that the statistical distribution for photons derived by Bose was far more general, and fundamental, than Bose himself had realized. In fact, in addition to translating Bose's paper into German, and forwarding it for publication, Einstein followed it up with a paper of his own, applying Bose's statistics to Helium nuclei. You may be aware that the helium nucleus consists of two neutrons and two protons. As you can see from Table 4.1, both neutrons and protons have spin equal to $\frac{1}{2}\hbar$. Recall my earlier statement that the spin angular momentum vector of a spin $-\frac{1}{2}\hbar$ particle can only point up or down (with respect to some chosen axis). In other words, the spin angular momentum can be either $+\frac{1}{2}$ or $-\frac{1}{2}$ (in units of $\hbar$). Convince yourself that whatever may be the spin orientation of the two protons and two neutrons inside the helium nucleus, the resultant spin of the four particles must be an *integral multiple of* $\hbar$. Therefore, *the helium nucleus must obey Bose's statistics*. Since it was Einstein who first realized this, one refers to the statistics as **Bose–Einstein statistics**.

This generalization by Einstein may seem rather straightforward, but it has profound consequence for the behaviour of liquid helium at extremely low temperatures, and this was pointed out by Einstein in 1925. In 2001, 76 years after Einstein made this prediction, three physicists were awarded the Nobel Prize for Physics for experimentally demonstrating the phenomenon which is now known as *Bose–Einstein Condensation*. We shall discuss some of this in the next book in this series, entitled, *Neutron Stars and Black Holes*, but we shall not digress to discuss it here. I refer you to the delightful book, *Bose And His Statistics,* by G. Venkataraman for a comprehensive and historical account.

Why do fermions and bosons obey different rules? What is the connection between *spin* and *statistics*? This is a deep question. Pauli showed that the connection between spin and statistics is to be found in relativistic quantum mechanics. Pauli's arguments are very involved and subtle. But no one has found a simple and straightforward answer to this basic question. Let us hear what Feynman has to say about this:

> ... It appears to be one of the few places in physics where there is a rule which can be stated simply, but for which no one has found a simple and easy explanation. The explanation is deep down in relativistic quantum mechanics. This probably means that we do not have a complete understanding of the fundamental principle involved. For the moment, you will just have to take it as one of the rules of the world.

This has been a long digression, but I hope you will find it useful to understand many of the things we shall be discussing in this series. What we have been discussing is also bread-and-butter stuff in modern condensed-matter physics.

After this review of the principles of statistical mechanics, let us now move on to a more detailed account of the Fermi–Dirac distribution.

Chapter 5
Fermi–Dirac Distribution

As discussed in the previous chapter, the probability distribution of particles like the electron, proton and neutron, with spin quantum number equal to [$\frac{1}{2}$], is given by the Fermi–Dirac distribution. This distribution was first invented for the electron by the Italian physicist **Enrico Fermi**. The relation of this distribution with quantum mechanics was elucidated by Paul Dirac in a seminal paper published in 1926. Hence, the names of both these great physicists are associated with this distribution (Fig. 5.1).

Pauli's Exclusion Principle

Wolfgang Pauli was the first to appreciate an important fact concerning spin $\frac{1}{2}$ particles like the electron. Let us consider two electrons in a box. Pauli noticed that *the wave function describing the two electrons must be antisymmetric in the coordinates and spins of identical particles*: that is, if the coordinates and spin of one particle are interchanged as a group with those of another, the wave function must merely change sign. In other words, we will pick up a *minus sign*. Another way of saying this is that wave function must be *antisymmetric* if we *exchange* the two particles. From this Pauli deduced a general rule which now bears his name:

No two electrons can be in the same electronic quantum state.

Fig. 5.1 Enrico Fermi (*left*) and Paul A. M. Dirac (*right*)

G. Srinivasan, *Life and Death of the Stars*, Undergraduate Lecture Notes in Physics, DOI: 10.1007/978-3-642-45384-7_5, © Springer-Verlag Berlin Heidelberg 2014

Put more simply, if there is already an electron in a quantum state then a second electron cannot occupy the same state. *Hence, for electrons, the occupation number of each state can only take the values 0 or 1*. Pauli enunciated this extremely important principle in 1925, before the discovery of wave mechanics by Schrödinger. Fermi and Dirac, quite independently, realized the profound significance of Pauli's Principle, and went on to construct the probability distribution that we shall now discuss.

The Fermi–Dirac Distribution

Let us consider a perfect gas of fermions, say a gas of electrons. Such a gas is usually referred to as a Fermi gas. The mean number of particles in the quantum state k, with energy E_k, is given by

$$\boxed{f(E_k) = \frac{1}{e^{(E_k-\mu)/kT} + 1}.} \tag{5.1}$$

This is the probability distribution of a perfect gas obeying Fermi–Dirac statistics (often referred to simply as the Fermi distribution). In the above expression, μ, is the chemical potential. In simple terms, μ is the energy needed to add one more particle to the system. Its meaning will become clearer soon. The total number of particles in the system, N, is obtained by adding up the mean number of particles in all the quantum states:

$$\boxed{\sum_k f(E_k) = \sum_k \frac{1}{e^{(E_k-\mu)/kT} + 1} = N.} \tag{5.2}$$

Equation (5.2) determines the chemical potential μ as an implicit function of T and N.

Fermi Gas of Elementary Particles

Consider a gas made up of elementary particles, say, an electron gas. The energy of an elementary particle is simply the kinetic energy of its motion.

$$E = \frac{1}{2m}(p_x^2 + p_y^2 + p_z^2). \tag{5.3}$$

The distribution function in this case is just over the phase space of the particles. You will remember that we introduced the notion of phase space when we were discussing the Maxwell–Boltzmann gas. It is a six-dimensional space, with three spatial dimensions (x, y, z) and three momentum dimensions (p_x, p_y, p_z). Hence the number of particles in an element of phase space $dp_x dp_y dp_z dV$ is obtained by multiplying the Fermi distribution of Eq. (5.1) by the density of states:

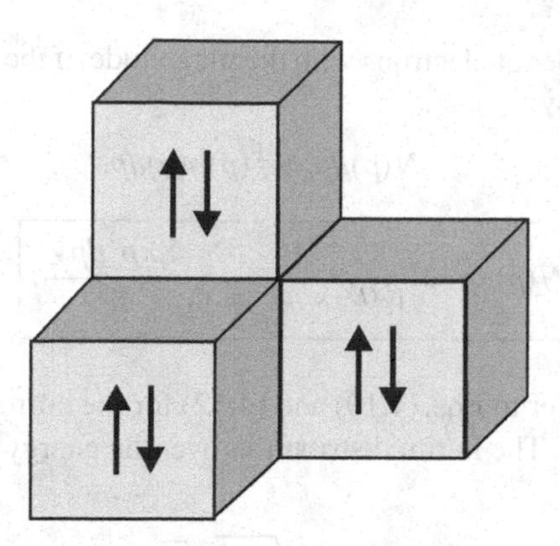

Fig. 5.2 To specify the quantum state of an electron, one has to specify not only its momentum (or energy) but also the direction of spin. The spin angular momentum vector of an electron can point in two opposite directions. Therefore, *two electrons with the same momentum but spins pointing in opposite directions represent two distinct quantum states*. Hence one can put two electrons in each cell in phase space, as shown in the figure. This would not be a violation of Pauli's exclusion principle

$$2 \times \frac{dp_x dp_y dp_z dV}{h^3} = 2 \times \frac{4\pi p^2 dp dV}{h^3}.$$

Remember that phase space is discrete since we are dealing with a quantum system. That is why the elementary volume in phase space $dp_x dp_y dp_z dV$ has been divided by the volume of the cells, namely, h^3. This will give us the number of cells in the desired momentum interval (Refer to Eq. (4.24) and Figs. 4.7 and 4.8). You will notice that the above equation differs from Eq. (4.24) by a factor, 2. The meaning of this should be clear from Fig. 5.2. *We are actually putting two electrons in each quantum cell in phase space.* Is this not a violation of Pauli's exclusion principle, according to which there can only be one electron in each quantum state? No, it need not be a violation for the following reason: Remember that in quantum mechanics an electron has *two* degrees of freedom: translational degree of freedom and spin degree of freedom. By specifying the momentum of the particle, we are defining the state of the electron as far as its translational degree of freedom is concerned. But to complete the description of the quantum state of the particle, we also have to specify the direction of the spin. Since the electron is a spin $\frac{1}{2}$ particle, the spin angular momentum vector can have two orientations, which we may call as *up* and *down*. These two orientations of the spin correspond to two distinct quantum states. Therefore, as long as we make sure that the two electrons have spins pointing in the opposite direction, we can put them in the same cell in phase space without violating Pauli's exclusion principle!

The distribution over the momentum of the particle is obtained by substituting the total volume V of the gas for dV. The meaning of such a substitution should be clear. If we are only interested in the magnitude of the momentum, the particles can obviously be *anywhere* inside the box.

Therefore, the number of electrons with the magnitude of the momentum between p and $p + dp$ is given by:

$$N(p)dp = f(p)g(p)dp,$$

$$\boxed{N(p)dp = \frac{1}{\left(e^{(E-\mu)/kT} + 1\right)} \frac{8\pi p^2 dp V}{h^3}.} \tag{5.4}$$

where $E = p^2/2m$. Refer to Eqs. (4.10) and (4.12) for the corresponding expression in Boltzmann statistics. The Fermi distribution over the energy can be written down by recalling Eq. (4.15)

$$p^2 dp = \sqrt{2m^3}\sqrt{E}dE,$$

$$\boxed{N(E)dE = \left(\frac{V8\pi\sqrt{2m^3}}{h^3}\right) \frac{1}{\left(e^{(E-\mu)/kT} + 1\right)} \sqrt{E}dE.} \tag{5.5}$$

Integrating Eq. (5.5) over the energy we obtain the total number of particles in the gas:

$$\boxed{N = \left(\frac{V8\pi\sqrt{2m^3}}{h^3}\right) \int_0^\infty \frac{1}{\left(e^{(E-\mu)/kT} + 1\right)} \sqrt{E}dE.} \tag{5.6}$$

The Degenerate Electron Gas

We shall now focus on the properties of an electron gas at *very low temperatures. The words low* or *high, big* or *small,* have no meaning in physics. These adjectives must be in comparison to something! A white dwarf star at a temperature of 10^5 K is a very cold object in the context of the present discussion. As we shall see in the next book in the series, a neutron star with an internal temperature of 10^7 K should be regarded as an incredibly cold object. The meaning of this will become clear presently.

An Electron Gas at Absolute Zero

The most dramatic way to bring out the difference between the Fermi distribution and the Boltzmann distribution is by considering the statistical properties of an electron gas at T = 0 K. In classical physics, all motions cease at absolute zero. This is natural because the motion of the particles is by virtue of heat; indeed, *heat* is just these random motions. The *internal energy of the gas is zero* at T = 0 K, *and the pressure of the gas also vanishes.*

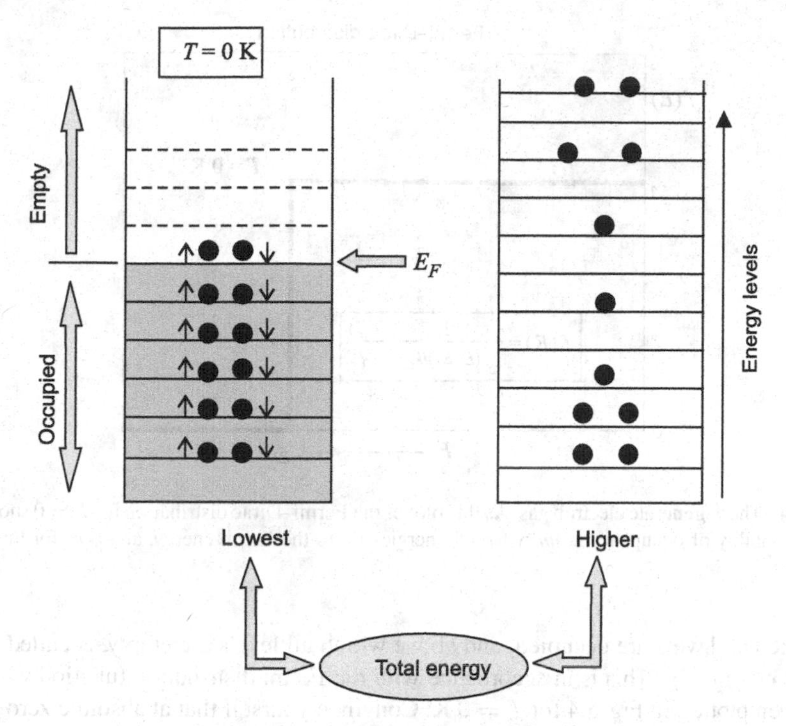

Fig. 5.3 *The degenerate electron gas*. In Fermi–Dirac statistics, an electron gas has finite energy even at the absolute zero of temperature. It has this energy by virtue of Pauli's exclusion principle, according to which there can be only one electron in a given quantum state. Therefore, *the occupation number of any state can only be zero or one*. One could, in principle, populate the energy levels as shown on the right-hand side of the figure. But this would not be the configuration with the lowest total energy, as is required at absolute zero. The arrangement on the left, where we put two electrons with opposite spins in each level, is clearly the lowest-energy configuration. The highest energy up to which all states are fully occupied is known as the *Fermi energy*

This is not so for an electron gas because it is a quantum gas. Since electrons have to obey Pauli's exclusion principle, all the electrons cannot be put in the zero energy state. *Pauli's principle demands that the occupation number of any state can only be* 0 or 1. Therefore, an electron gas will have finite energy even at absolute zero of temperature! This energy has nothing to do with heat. The internal energy of the gas is a consequence of Pauli's exclusion principle. This is schematically shown in Fig. 5.3.

Having settled that, let us ask *how* the electrons will distribute themselves amongst these energy levels. This is also shown in Fig. 5.3. The fundamental principle here is that *the electrons will distribute themselves in such a way as to minimize the total energy*. The lowest total energy configuration is shown on the left. The first thing is to exploit the fact that we can, in fact, put two electrons in each state, *provided they have opposite spins*. This does not violate Pauli's rule. So, starting from the lowest energy level, we put two electrons in each state, till we run out of electrons. The largest value of the energy up to which all the levels are full will, obviously, be determined by the total number of electrons in the gas. This maximum energy up

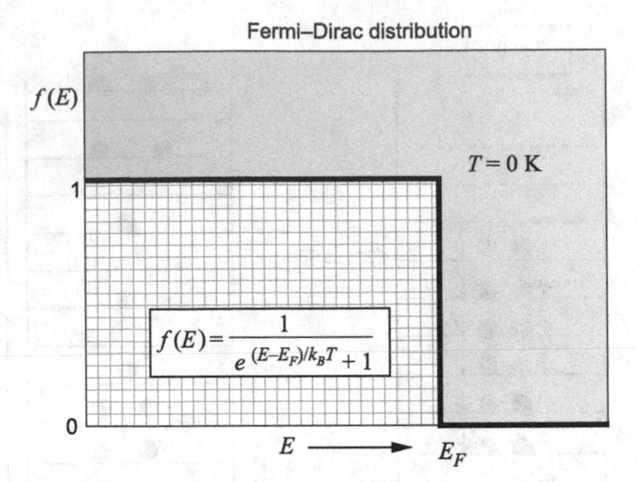

Fig. 5.4 The degenerate electron gas. As this plot of the Fermi–Dirac distribution for $T = 0$ shows, the probability of occupancy is *unity* for all energies up to the Fermi energy, and *zero* for higher energies

to which all levels are occupied, and above which all levels are empty, is called the *Fermi energy,* E_F. This is in accordance with the Fermi distribution function which has been plotted in Fig. 5.4 for $T = 0$ K. Convince yourself that at absolute zero the probability of occupancy $f(E)$ is unity for all states with $E < E_F$ and zero for all states with $E > E_F$.

Let us now derive the expression for the Fermi momentum and Fermi energy of a degenerate gas of N electrons in a volume V.

Fermi Momentum

It is convenient to go back to momentum space and determine what is known as Fermi momentum, p_F, which is related to the Fermi energy by $E_F = p_F^2/2m$.

Let us start with the cell at the origin in Fig. 5.5 and put two electrons in it with opposite spin. Let us then move outward in a systematic manner, in all three directions, putting two electrons in each cell. At some stage, we shall run out of electrons. Let p_F be the radius of the sphere that defines the outer envelope of the occupied cells. You might wonder how you can get a sphere by stacking cubes! Well, if the number of cubes is very large, or if the cubes are small, then the stack of cubes will approximate a sphere to a very good accuracy.

The radius of the sphere, p_F, inside which all the cells are occupied, and outside which all the cells are empty is known as the *Fermi momentum* (refer to Fig. 5.5). Given a volume V, the Fermi momentum is determined by the total number of particles N. The number of cells inside the sphere will be equal to the volume of the sphere divided by the volume of the quantum cells.

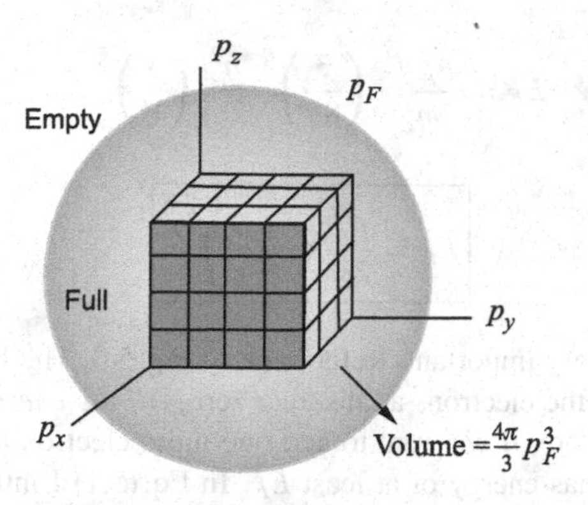

Fig. 5.5 This figure explains the concept of *Fermi momentum*. Given N electrons, the way to populate the cells in phase space is to start with the cell at the origin and move outwards, putting two electrons in each cell, until one has run out of electrons. The radius of the sphere circumscribing the filled cells is known as the Fermi momentum p_F. The physical significance of p_F is that this is the highest momentum of the electrons. As may be seen from Eq. (5.7) and (5.8), the Fermi momentum is proportional to the one-third power of the electron density

$$\text{Number of occupied cells} = \frac{\left(\frac{4\pi}{3}p_F^3\right)}{\left(\frac{h^3}{V}\right)}.$$

(Refer to Fig. 4.7). Since we are putting two electrons in each cell,

$$\boxed{N = 2 \times \text{number of occupied cells} = \frac{2V}{h^3}\left(\frac{4\pi}{3}p_F^3\right).} \tag{5.7}$$

Simplifying Eq. (5.7) we get for the Fermi momentum:

$$\boxed{p_F = \left(\frac{3}{8\pi}\right)^{\frac{1}{3}} h \left(\frac{N}{V}\right)^{\frac{1}{3}}.} \tag{5.8}$$

The Fermi momentum is thus proportional to the one-third power of the number density of electrons:

$$\boxed{p_F \propto \left(\frac{N}{V}\right)^{\frac{1}{3}}} \tag{5.9}$$

Fermi Energy

The Fermi energy can now be easily determined. Using Eq. (5.8), we get:

$$E_F = \frac{p_F^2}{2m} = \left(\frac{3}{8\pi}\right)^{\frac{2}{3}} \frac{h^2}{2m} \left(\frac{N}{V}\right)^{\frac{2}{3}}, \tag{5.10}$$

$$\boxed{E_F = \frac{p_F^2}{2m} \propto \left(\frac{N}{V}\right)^{\frac{2}{3}}.} \tag{5.11}$$

The above result is very important. Refer back to Fig. 5.3. The Fermi energy is the maximum energy of the electrons at absolute zero. *All the energy levels up to that value are full.* Therefore, if we want to add one more electron to the box then one can only do so if it has energy of at least E_F. In Eq. (5.1) I introduced the notion of the chemical potential μ, which is the energy needed to add one more particle to the system. We now see that the chemical potential of a Fermi gas at absolute zero coincides with the Fermi energy.

$$\mu = E_F. \tag{5.12}$$

The important feature of Eqs. (5.9) and (5.11) is illustrated in Fig. 5.6. Given the total number of electrons, both the Fermi momentum and the Fermi energy will increase if we decrease the volume ($p_F \propto n^{\frac{1}{3}}$; $E_F \propto n^{\frac{2}{3}}$; $n = N/V$). Let us try to understand this via the *uncertainty principle*. You will recall from our discussion in Chap. 4, 'The Principles of Statistical Mechanics', that the discreteness of momentum space was a direct consequence of the uncertainty principle. The length, breadth and height of the elementary cells in phase space was determined by the uncertainty in the three components of the momentum, which, in turn, is determined by the size of the box ($\Delta p_x \sim h/L$, etc.). Consequently, the volume of the quantum cells is given by:

$$\text{Volume of cell} = \left(\frac{h}{L}\right)^3 = \frac{h}{V},$$

see Fig. 4.7). It is therefore easy to see why the size of the Fermi sphere in Fig. 5.6 increases with decreasing volume. The increase in the Fermi energy is a direct consequence of this.

Ground State Energy of a Degenerate Electron Gas

It should be clear from the above discussion *that an electron gas has energy even at absolute zero of temperature*. This energy is often referred to as *ground state energy* or *zero point energy*. Let us now calculate this energy.

$$E_{\text{Total}} = \int_0^\infty Ef(E)g(E)dE. \tag{5.13}$$

You will recall that $f(E)$ is the probability that the state with energy E is occupied, and $g(E)dE$ is the number of levels with energy in the range E to $E + dE$. Look at this equation in the following manner. Imagine that you own a skyscraper apartment

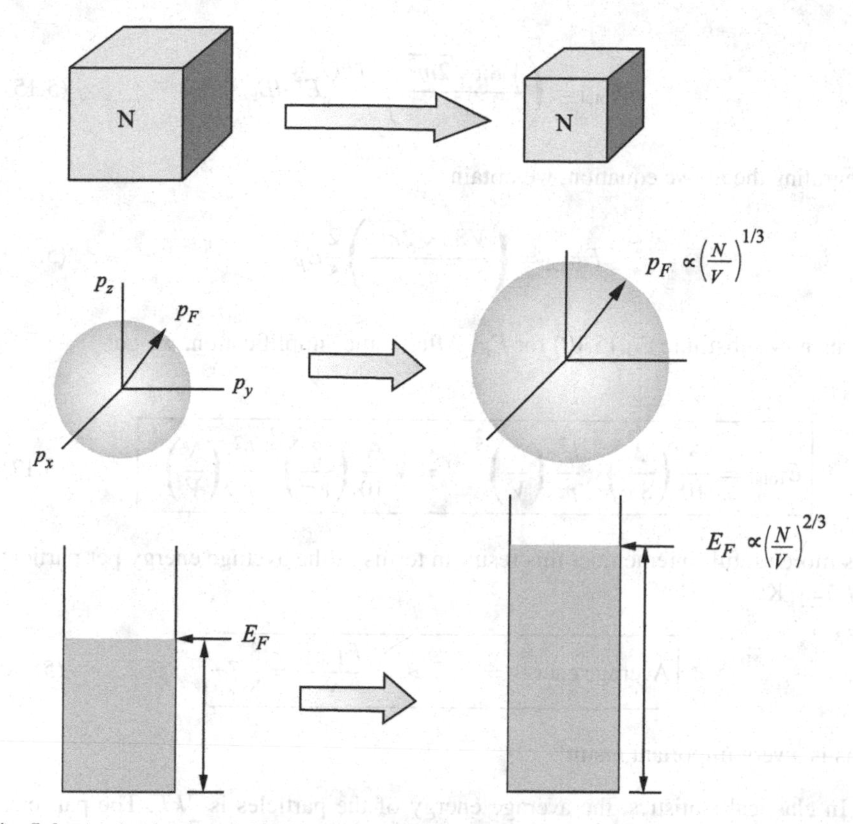

Fig. 5.6 As the density of the electron gas increases, both the Fermi momentum and the Fermi energy increase. As may be seen from Eq. (5.18) the average energy of the electrons at absolute zero is $\frac{3}{5}E_F$. Therefore the total energy, equal to the average energy times the total number of electrons, also increases with increasing density. As a consequence, one can compress an electron gas only by virtue of giving it enormous energy. Since this energy is an inevitable consequence of Pauli's exclusion principle, it represents *ground state energy*; *the electron gas cannot have less energy than this!*

building. The rent for the apartment increases as you go to higher floors, since the view is better. *The total rent you will collect from the building is equal to the sum of the rent from each floor. The rent from each floor is equal to the product of the rent per apartment on a given floor, multiplied by the probability that the apartment is occupied, multiplied by the number of apartments on that floor!* This is exactly the prescription given in Eq. (5.13) for calculating the total energy of the gas. We can now use Eq. (5.5) to evaluate the total energy.

$$E_{Total} = \left(\frac{V8\pi\sqrt{2m^3}}{h^3}\right)\int_0^\infty E\frac{1}{\left(e^{(E-\mu)/kT}+1\right)}\sqrt{E}dE \tag{5.14}$$

At absolute zero, the probability of occupancy is 1 for $E < E_F$ and zero for $E > E_F$ (Fig. 5.4). Therefore, the upper limit of the integral in Eq. (5.14) can be replaced by E_F. Therefore,

$$E_{\text{Total}} = \left(\frac{V8\pi\sqrt{2m^3}}{h^3} \right) \int_0^{E_F} E^{\frac{3}{2}} dE \qquad (5.15)$$

Integrating the above equation, we obtain

$$E_{\text{Total}} = \left(\frac{V8\pi\sqrt{2m^3}}{h^3} \right) \frac{2}{5} E_F^{\frac{5}{2}} \qquad (5.16)$$

Let us now substitute Eq. (5.10) for E_F. After some simplification, we get

$$\boxed{E_{\text{Total}} = \frac{3}{10} \left(\frac{3}{8\pi} \right)^{\frac{2}{3}} \frac{h^2}{m} \left(\frac{N}{V} \right)^{\frac{2}{3}} N = V \frac{3}{10} \left(\frac{3}{8\pi} \right)^{\frac{2}{3}} \frac{h^2}{m} \left(\frac{N}{V} \right)^{\frac{5}{3}}} \qquad (5.17)$$

It is more useful to remember this result in terms of the average energy per particle at $T = 0$ K:

$$\boxed{\text{Average energy} = < E > = \frac{E_{\text{Total}}}{N} = \frac{3}{5} E_F} \qquad (5.18)$$

This is a very important result.

1. In classical statistics, the average energy of the particles is $\frac{3}{2}kT$. The particles have no energy at absolute zero.
2. But in Fermi–Dirac statistics, the average energy of the particles is $\frac{3}{5}E_F$.
3. The denser the electron gas, the greater is this average energy since $E_F \propto n^{\frac{2}{3}}$.

Degeneracy Pressure

In classical statistical mechanics, the pressure of an ideal gas is related to its temperature. According to Boyle's Law, $P = nkT$. It is obvious that this pressure tends to zero as we approach absolute zero. Let us now calculate the pressure of a degenerate electron gas at absolute zero. We expect it to have a nonzero pressure even at absolute zero since it has internal energy. According to thermodynamics, pressure is related to the internal energy by the relation:

$$P = \frac{2}{3} \frac{E_{int}}{V}. \qquad (5.19)$$

If you are not familiar with this result, let us derive Boyle's Law using this relation. The total internal energy of an ideal classical gas is given by

$$E_{\mathrm{int}} = N \times \text{average energy of particles} = N \times \frac{3}{2}kT.$$

Therefore, $\frac{2}{3}\frac{E_{\mathrm{int}}}{V} = nkT$, which is the expression for pressure according to Boyle's Law.

The pressure of a degenerate electron gas can be obtained from Eq. (5.17). Using the second of the expressions on the right-hand side, we get

$$\boxed{P_{\mathrm{deg}} = \frac{2}{3}\frac{E_{\mathrm{Total}}}{V} = \frac{1}{5}\left(\frac{3}{8\pi}\right)^{\frac{2}{3}}\frac{h^2}{m}\left(\frac{N}{V}\right)^{\frac{5}{3}} \propto \left(\frac{N}{V}\right)^{\frac{5}{3}}} \tag{5.20}$$

Notice some of the important features of the above expression for the pressure of a Fermi gas.

1. A Fermi gas exerts nonzero pressure even at absolute zero.
2. The degeneracy pressure is $\propto n^{\frac{5}{3}}$.
3. The mass of the fermion enters the denominator of Eq. (5.20). Since the proton (or neutron) is roughly two-thousand times heavier than the electron, *the degeneracy pressure of a neutron gas or a proton gas will be about two thousand times less than the pressure of the electron gas*, even though the number density of neutron or protons might be the same as that of the electrons.

Fermi Gas at Finite Temperature

So far we have discussed the properties of an electron gas at absolute zero. We did this to bring out the essential difference between classical and quantum statistics. But in real physical situations, the electron gas will be at a finite temperature. Consider a metal like copper. It is a very good conductor because there are a lot of free electrons that are not tied to individual nuclei. When we bring atoms together, copper atoms in this case, the outermost electrons become unbound. These electrons are free to wander around the entire volume of the metal. We thus have an electron gas. Are we to describe these electrons using Fermi–Dirac distribution or Boltzmann distribution?

In order to answer this question, let us perturb the zero-temperature Fermi distribution shown in Fig. 5.7 by heating the electron gas. The description we have given of a strongly degenerate Fermi gas can be used as an excellent approximation for temperatures sufficiently close to absolute zero. The condition that this description should be applicable at finite temperatures requires that the *thermal energy kT* be very small compared to the *Fermi energy* E_F:

$$kT \ll E_F \tag{5.21}$$

Let us go back to our discussion of the free electron gas in copper at room temperature ~300 K. The Fermi energy of the electrons can be calculated using Eq. (5.10).

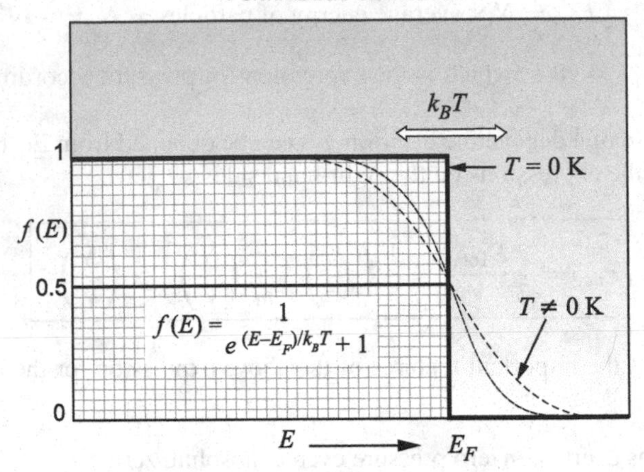

Fig. 5.7 The figure shows the Fermi–Dirac distribution at finite temperature. Since $kT \ll E_F$, the zero-temperature distribution (Fig. 5.4) is modified only slightly. A few electrons just below the Fermi energy move to higher energies, and the probability distribution develops a *tail* whose width is roughly kT. Below a depth of kT, the electron distribution remains unaltered—the thermal energy is not enough to alter the distribution below this depth. When $kT \gg E_F$, the *tail* is fully developed and we shall recover the Boltzmann distribution

A simple calculation gives $E_F\sim$ a few electron volts. Now, 1 eV is approximately equal to 10^4 kelvin, in temperature units. To put it differently,

$$\frac{1 \text{ eV}}{k_B} \approx 10^4 \text{ K}$$

Clearly, the thermal energy kT is much smaller than the Fermi energy E_F. Therefore, the electron gas in copper should be regarded as strongly degenerate at room temperature.

As we increase the temperature of the gas, the Fermi distribution develops a *tail* whose width is $\sim kT$ (see Fig. 5.7). A few electrons originally below the Fermi energy can spill over. Since $kT \ll E_F$ it is only the electrons with energy already close to the Fermi energy that can spill over. As the temperature is increased further, the tail of the Fermi distribution will become a little more pronounced. The temperature determined by the relation $kT_0 \cong E_F$ is often referred to as the *degeneracy temperature*. This is roughly the temperature at which quantum effects begin to become important. Notice that this is not a fixed temperature. It depends on the electron density, since the Fermi energy depends on the electron density.

At $T \gg T_0$, the Fermi distribution will transform to the familiar Boltzmann distribution. This is because, in a dilute gas, at sufficiently high temperatures, Pauli's exclusion principle is unlikely to have any observable consequences.

We are now ready to discuss R. H. Fowler's historic resolution of Eddington's paradox concerning the companion of Sirius.

Chapter 6
Quantum Stars

Fowler to the Rescue of White Dwarfs

Let us now pick up from where we left off at the end of Chap. 3, 'White Dwarf Stars'. We were discussing the strange companion of Sirius, *a star with a mean density approximately equal to a million times the density of the Sun*. You will recall that Eddington was concerned what will happen to such ultra-dense stars when their *supply of subatomic energy fails*. He famously stated, 'The star will need energy in order to cool'. **Sir Ralph Howard Fowler**, shown in Fig. 6.1, a colleague of Eddington and a professor of theoretical physics at Cambridge University stated Eddington's paradox thus: 'The stellar material will have radiated so much energy that it has less energy than the same matter in normal atoms expanded at the absolute zero of temperature. If part of it were removed from the star and the pressure taken off, what could it do?'

We argued in Chap. 3 that if the pressure is released, the stellar matter can resume its original state of existence as a collection of normal atoms only if the kinetic energy per unit volume is greater than the attractive electrostatic energy per unit volume. We concluded that this condition would be met only if

$$\rho < (0.94 \times 10^{-3} T/\mu Z^2)^3.$$

At densities and temperatures expected in the interior of the white dwarf, the kinetic energy per unit volume will, in fact, *be less than* the electrostatic energy:

$$E_K(\text{perfect gas}) < E_V.$$

Consequently, the matter will not be able to resume the state of being a collection of ordinary atoms, and the star will, indeed, be in an awkward predicament when its energy supply fails, as Eddington had feared. This is, of course, assuming that the stellar matter behaved as a *perfect gas*.

G. Srinivasan, *Life and Death of the Stars*, Undergraduate Lecture Notes in Physics,
DOI: 10.1007/978-3-642-45384-7_6,

Fig. 6.1 Sir Ralph Howard Fowler

R. H. Fowler resolved this paradox in 1926 by invoking what was at the time *hot-off-the-press* statistical mechanics of Fermi and Dirac. This paper by Fowler is one of the great landmarks in the development of our ideas concerning stellar structure and stellar evolution. What is equally remarkable was the speed with which Fowler had absorbed the new development and applied it to resolve the above paradox. Dirac was one of Fowler's students when he derived the statistical distribution we introduced in Chap. 5, 'Fermi–Dirac Distribution'. Fowler communicated Dirac's paper to the Royal Society on 26 August 1926. On 3 November, Fowler communicated a paper of his own to the Royal Society in which he applied the new statistics to an assembly of identical particles; in other words, this was much of what we described in Chap. 5. On 10 December, Fowler presented a paper entitled, 'Dense matter', before the Royal Astronomical Society. In this historic paper Fowler drew attention to the fact that *the electron gas in matter that was as dense as in the companion of Sirius must be degenerate* (in the sense in which we explained in the previous chapter).

Thus, Fowler was the very first person to *apply* the new statistics of Fermi and Dirac. Extraordinarily, the first application of a new *quantum principle* was to a *star*! Soon after that, Pauli invoked the Fermi–Dirac statistics to explain the *paramagnetism* of the alkali metals. This was followed by a classic paper by the great German physicist **Arnold Sommerfeld** in which he developed what has come to be known as the *Free Electron Theory of Metals*. (Sommerfeld had a reputation as a great teacher and attracted many brilliant young men. Among his students from that epoch are

Wolfgang Pauli, Peter Debye, Werner Heisenberg, Gregor Wenzel, Hans Bethe, Rudolf Peierls and others! A list that is unmatched in the history of science.)

Fowler's resolution of Eddington's paradox was simply this: *since the electrons will be degenerate at the densities and temperatures in the white dwarfs, the kinetic energy, E_K,per unit volume should be evaluated using the Fermi–Dirac statistics and not using Boyle's law. He showed that when E_K is so evaluated, it is indeed much greater than E_V.*

$$E_K(\text{Fermi–Dirac}) \gg E_V$$

So, if the pressure is taken off, the stellar material will be able to assume its original state of normal atoms. The white dwarfs need not worry! When their supply of heat is exhausted, Pauli's Exclusion Principle and Fermi–Dirac statistics would ensure that they would die a peaceful death.

Enter Chandra

The year 1928 was a momentous one in the history of Indian science. In February that year, **C. V. Raman** and his student **K. S. Krishnan** discovered an important effect which has come to be known as the *Raman Effect*. Raman's nephew, **Subrahmanyan Chandrasekhar**, shown in Fig. 6.2, was a first year B.Sc. Honours student at Presidency College, Madras (now Chennai). That summer, after completing his first year, Chandra (as he came to be known the world over) went to Calcutta (now Kolkata), to visit Raman and his young students at the Indian Association for the Cultivation of Science.

The place was buzzing with excitement. Raman was in a state of euphoria. There was expectation that Raman would be awarded the Nobel Prize for this important discovery he had made. This came true in 1930, when Raman was awarded the Nobel Prize for Physics. Young Chandra must have felt inspired to be in such an environment.

In the fall of 1928, Arnold Sommerfeld visited Madras and delivered a lecture at the Presidency College. Chandra was terribly excited, not only because Sommerfeld was one of the greatest physicists in the world, but also because he had read Sommerfeld's book, *Atomic Structure and Spectral Lines*. He took an appointment and went to meet Sommerfeld at his hotel. It was during this conversation that Chandra learnt of the great transformation that was taking place in physics: the discovery of wave mechanics by Schrödinger, and the new developments due to Heisenberg, Dirac, Pauli and others. Sommerfeld also talked about the new developments in statistical mechanics due to Fermi and Dirac. In fact, he gave Chandra a copy of his unpublished paper, 'The free electron theory of metals'.

This meeting with Sommerfeld had a great impact on Chandra's evolution as a physicist. He carefully studied the paper that Sommerfeld had given him. This paper was in German, but this posed no difficulty since, like most serious students of

Fig. 6.2 Subrahmanyan Chandrasekhar

physics at that time, Chandra was well versed in German. Thus he learnt about the new statistics of Fermi and Dirac. He immersed himself in the University Library and frantically acquainted himself with the new developments in physics. Among other things, he came across the paper by Fowler that we have referred to. Within a few months, he had written a paper entitled, 'Compton scattering and the new statistics'. Having written this paper, he had the audacity to send it to Fowler in Cambridge, requesting him to communicate it to the Royal Society! Fowler was sufficiently impressed with the paper that he got it published in the journal, *Proceeding of the Royal Society.* Chandra was barely eighteen years old at that time!

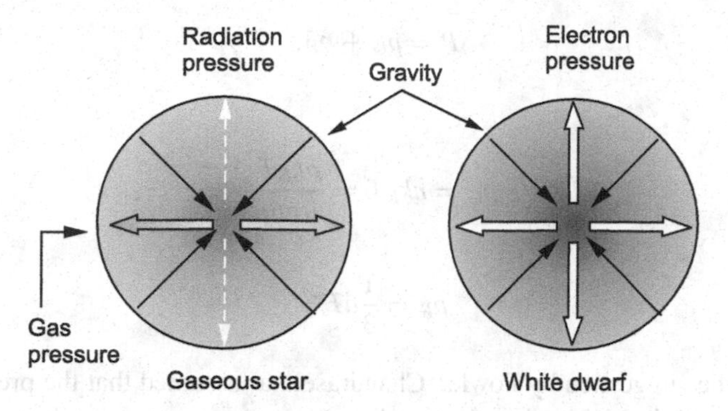

Fig. 6.3 In a diffused star like the Sun, the inward pull due to gravity is balanced by the combined effect of the pressure of ideal gas and radiation pressure. While the gas pressure depends on both density and temperature, radiation pressure depends upon temperature alone. The relative proportion of gas pressure and radiation pressure depends upon the mass of the star, with radiation pressure becoming increasingly important in more massive stars. Fowler's brilliant idea was that in a white dwarf gravity is balanced by the pressure of the electrons that arises due to Pauli's exclusion principle. Since this pressure exists even at absolute zero, a white dwarf can be stable even at absolute zero!

Chandrasekhar's Theory of the White Dwarfs

Chandra did not rest on his laurels. He was inspired by Fowler's paper on the dense stars and went on to construct a proper theory of the white dwarfs. He had read Eddington's book, *The Internal Constitution of the Stars,* and mastered all the mathematical tools needed to do this. Let us now discuss the main new results obtained by him.

Since Chandra's starting point was the remarkable paper by Fowler, let us briefly recall what Fowler had suggested. The essence of Fowler's prescient suggestion is schematically shown in Fig. 6.3.

You will recall that in Eddington's theory the stellar material is assumed to be an ideal gas. The inward-directed gravity is balanced by the combined effect of gas pressure and radiation pressure. *Fowler's idea was that in a highly dense star such as a white dwarf, gravity is balanced by the degeneracy pressure of the electrons.*

One thing that Chandra set out to do was to obtain the relationship between the mass and the radius of a white dwarf. To derive the mass–radius relation, one must integrate the *equation of hydrostatic equilibrium*, with an assumed equation of state (that is, an expression for the pressure in terms of the density). We discussed this equation in Chap. 1 of ***What Are the stars?***, but let us recall it once again:

$$\boxed{\frac{dP}{dr} = -\frac{GM(r)\rho(r)}{r^2}.} \tag{6.1}$$

In Eddington's theory, the pressure on the left-hand side of the above equation is the sum of gas pressure and radiation pressure:

$$P = p_G + p_R, \tag{6.2}$$

where,

$$p_G = nk_BT = \frac{\rho k_B T}{\mu m_H} \tag{6.3}$$

$$p_R = \frac{1}{3}aT^4. \tag{6.4}$$

Following the suggestion by Fowler, Chandrasekhar assumed that the pressure that balances gravity in a white dwarf is the *degeneracy pressure of the electrons*:

$$\boxed{P_{\text{deg}} = \frac{2}{3}\frac{E_{\text{Total}}}{V} = \frac{1}{5}\left(\frac{3}{8\pi}\right)^{\frac{2}{3}}\frac{h^2}{m_e}\left(\frac{N}{V}\right)^{\frac{5}{3}}.} \tag{6.5}$$

Here, N is the number of *electrons* in volume V. It would be more useful to follow Eq. (6.3) for Boyle's law and rewrite the degeneracy pressure as a function of the mass density ρ, instead of the number density $n = (N/V)$.

If we were dealing with a gas of protons, say, then $\rho = n_p m_p$, where n_p is the number density of protons, and m_p is the mass of the proton. On the other hand, if the gas is ionized hydrogen then there will be equal number of electrons and protons, with $n_e = n_p$. But the mass of the electron is negligible when compared with the mass of the proton. Therefore, we should keep this in mind when we convert the number density of electrons into mass density. Since the mass is essentially determined by the number of protons, $\rho = n_p m_p$. However, since $n_e = n_p$, it follows, $\rho = n_p m_p = n_e m_p$, or $n_e = \rho/m_p$.

Things will be slightly more complicated if the gas is an admixture of hydrogen, helium, carbon, etc. Each atom of helium, for example, will contribute *two electrons* and *four nucleons* (neutrons plus protons) to the stellar plasma. Therefore, the number density of electrons will be one-half the number density of nucleons:

$$n_e = \frac{1}{2}(n_n + n_p) = \frac{1}{2}n_{\text{nucleons}}. \tag{6.6}$$

Multiplying and dividing the right-hand side by m_p, we get

$$n_e = \frac{n_{\text{nucleon}} m_p}{2m_p} = \frac{\rho}{2m_p}. \tag{6.7}$$

Instead of helium, let us consider some other heavy element (A, Z), where A is the number of nucleons and Z is the number of electrons. In this case,

$$\frac{n_e}{n_{\text{nucleon}}} = \frac{Z}{A}.$$

Hence, Eq. (6.7) can be written in the more general form:

$$n_e = \frac{Z}{A} n_{\text{nucleon}} = \frac{Z}{A} \frac{n_{\text{nucleon}} m_p}{m_p} = \frac{\rho}{\frac{A}{Z} m_p} \simeq \frac{\rho}{2m_p}. \tag{6.8}$$

You will notice that in Eq. (6.8) we have substituted 2 for (A/Z) in the denominator. If you refer to the famous *Periodic Table*, you will discover that for all elements barring hydrogen, the number of protons inside the nucleus is very nearly equal to the number of neutrons. In other words, $(A/Z) \approx 2$ for all elements (except hydrogen). In reality, one does not have pure hydrogen or helium or carbon or oxygen, etc. One has an admixture of various elements. One then introduces the concept of *mean molecular weight per electron*, μ_e, and writes (6.8) in the generalized form:

$$\boxed{n_e = \frac{\rho}{\mu_e m_p}.} \tag{6.9}$$

It is an interesting to note that as long as there is no hydrogen, $\mu_e \approx 2$, regardless of the relative abundance of elements. As we shall see in later chapters, by the time a star ends its life as a white dwarf, it would have consumed all the hydrogen in the core. And it is the *core* that then becomes a white dwarf. Thus, $\mu_e \approx 2$ is a good assumption. Do not be confused by the term, *mean molecular weight.* It is a misnomer! It should be clear from Eq. (6.9) that $\mu_e m_p$is the *average mass per particle.* To summarise the above discussion, if we want to convert the number density of electrons to the mass density, the prescription is:

$$\boxed{n_e = \frac{\rho}{2m_p}.}$$

After this digression, let us go back to Eq. (6.5) and rewrite the electron degeneracy pressure in terms of the mass density:

$$\boxed{P_{\text{deg}} = K_1 \rho^{\frac{5}{3}},} \tag{6.10}$$

where

$$\boxed{K_1 = \frac{1}{5}\left(\frac{3}{8\pi}\right)^{\frac{2}{3}} \frac{h^2}{m_e} \frac{1}{\left(\mu_e m_p\right)^{\frac{5}{3}}}.} \tag{6.11}$$

Mass–Radius Relation

To model a white dwarf of a certain mass we have to solve the equation of hydrostatic equilibrium, Eq. (6.2), supplemented by the equation of state (6.10):

$$\frac{dP}{dr} = -\frac{GM(r)\rho(r)}{r^2},$$

$$P_{\text{deg}} = K_1 \rho^{\frac{5}{3}}.$$

This is an easier problem than the one Eddington had to solve. In the case of gaseous stars, the total pressure is a function of both density and temperature, see Eqs. (6.3) and (6.4). Therefore, the density gradient in the star, and the temperature gradient, are interrelated. In the case of a white dwarf, the degeneracy pressure is a function of density only. Strictly speaking, this is true only at absolute zero. You will say that the companion of Sirius is a very hot star. That is true! But as I remarked in Chap. 5, The Fermi–Dirac distribution, under the temperature and density that are found in a white dwarf, $kT \ll E_F$, and our zero-temperature approximation to the white dwarf is an extremely good assumption.

There is an elaborate mathematical machinery to solve an equation of the type (6.2), when the pressure is related to the density as follows:

$$P = K\rho^{1+\frac{1}{n}}. \tag{6.12}$$

Such an equation of state is known as a polytrope of index n. In our case, $n = \frac{3}{2}$. Using this standard machinery, Chandrasekhar derived the following relationship between the radius and mass of a white dwarf:

$$\boxed{R = \left(\frac{K_1}{0.424G}\right)\frac{1}{M^{\frac{1}{3}}}}, \tag{6.13}$$

$$\boxed{R \propto M^{-\frac{1}{3}}.}$$

In Eq. (6.13), K_1 is the constant defined by (6.11) and G is Newton's constant of gravity. Equation (6.13) is the famous mass–radius relation for white dwarfs, derived by Chandrasekhar in 1929: *The radius of a white dwarf is inversely proportional to the cube root of the mass*. This is graphically shown in Fig. 6.4.

Let us now try to derive this using a heuristic derivation. The equation we want to solve is:

$$\frac{dP}{dr} = -\frac{GM(r)\rho(r)}{r^2}.$$

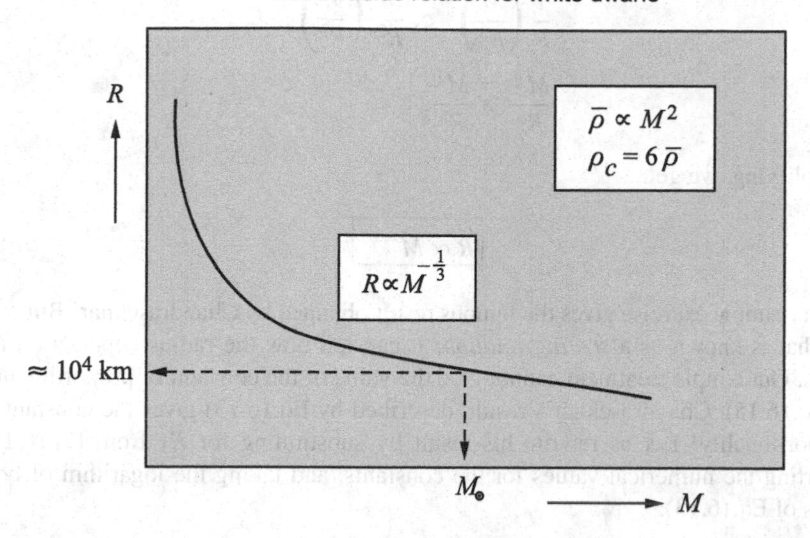

Fig. 6.4 The mass–radius relation for white dwarfs derived by Chandrasekhar in 1929. The radius is *inversely proportional* to the cube root of the mass of the white dwarf. This is in contrast to terrestrial matter, planets, etc. for which the radius will increase as the mass increases. A white dwarf with a mass equal to the mass of the Sun will have a radius roughly equal to the radius of the Earth! Remember that the radius of the Sun is roughly a million kilometres. Consequently, the mean density of a white dwarf of one solar mass will be roughly a million grams per cubic centimetre

Let us replace the *differential* by a *difference*:

$$\frac{dP}{dr} \approx \frac{(P(r) - P_{\text{surface}})}{R} = \frac{P}{R}.$$

Here, R is the radius of the star. Remember that the pressure at the *surface* is zero. With this approximation, the equation of hydrostatic equilibrium reads:

$$\frac{P}{R} \propto \frac{M\rho}{R^2}. \tag{6.14}$$

(In the simple-minded discussion below, we shall not display the fundamental constants, numerical constants, etc.) Let us now use the equation of state

$$P \propto \rho^{\frac{5}{3}},$$

and $\rho \propto \frac{M}{R^3}$. With these substitutions, Eq. (6.14) gives

$$\frac{1}{R}\left(\frac{M}{R^3}\right)^{\frac{5}{3}} \propto \frac{M}{R^2}\left(\frac{M}{R^3}\right),$$

$$\frac{M^{\frac{5}{3}}}{R^6} \propto \frac{M^2}{R^5}.$$

Simplifying, we get:

$$\boxed{R \propto M^{-\frac{1}{3}}.} \tag{6.15}$$

So our simple exercise gives the famous result obtained by Chandrasekhar! But ours is what is known as a *scaling relation;* it can tell how the radius depends on the mass. Our simple treatment cannot give the value of the constant of proportionality in Eq. (6.15). Chandrasekhar's result, described by Eq. (6.13) gives the constant of proportionality! Let us rewrite his result by substituting for K_1 from Eq. (6.11), inserting the numerical values for the constants, and taking the logarithm of both sides of Eq. (6.13):

$$\boxed{\log_{10}\left(\frac{R}{R_\odot}\right) = -\frac{1}{3}\log_{10}\left(\frac{M}{M_\odot}\right) - \frac{5}{3}\log_{10}\mu_e - 1.397.} \tag{6.16}$$

For a mass equal to the solar mass and $\mu_e = 2$, Eq. (6.16) predicts $R = 1.26\times10^{-2}R_\odot$ (approximately 10,000 km) and a mean density of 7.0×10^5 g cm^{-3}. These values are precisely of the order of the radii and mean densities encountered in white dwarf stars, such as the companion of Sirius. I am sure you are intrigued by an important feature of Fig. 6.3. In our experience, as the mass increases, the size of the object *increases*. It is exactly the opposite for a white dwarf! Think about this.

Besides the mass–radius relation, Chandrasekhar derived two other important results from his theory, which we have summarized below.

1. *The radius of a white dwarf is inversely proportional to the cube root of the mass.*
2. *The mean density is proportional to the square of the mass.*
3. *The central density is six times the mean density.*

Before proceeding further, we should ask the following question. Why was it that Fowler and Chandrasekhar considered only the pressure of the electrons? After all, the stellar plasma consists of an equal number of protons and roughly an equal number of neutrons. Since neutrons and protons are Fermions they, too, obey the Fermi–Dirac statistics. Therefore, should we not consider the degeneracy pressure of the nucleons also? This should bother you, since in a classical gas obeying Boyles' law, all the species of particles contribute to the pressure in equal measure, if their number densities are the same. This is because of the *equipartition theorem* according to which the average energy of the particles is $\frac{3}{2}kT$, regardless of their mass; a proton or a neutron have the same average energy as an electron (their speeds will, of course, be different). Therefore all the species of particles contribute equally to the internal energy and the pressure.

But this is not so for a Fermi gas. If you refer back to Eq. (5.20) which we have reproduced for ready reference, you will see that the mass of the particle enters the denominator of the expression for the degeneracy pressure.

$$P_{\text{deg}} = \frac{2}{3}\frac{E_{\text{Total}}}{V} = \frac{1}{5}\left(\frac{3}{8\pi}\right)^{\frac{2}{3}}\frac{h^2}{m}\left(\frac{N}{V}\right)^{\frac{5}{3}}.$$

Therefore the quantum pressure of the neutrons and protons is roughly two-thousand times less than that of the electrons, even if their number densities are the same.

Quite apart from that, one should not assume that the nucleons should necessarily be described by quantum statistics. If you recall our discussion in Chap. 5, the condition for a Fermi gas to regarded as degenerate is that the *thermal energy* kT is very small compared to the *Fermi energy* E_F:

$$kT \ll E_F.$$

While this condition would be satisfied for the electron gas, this inequality may not hold for the nucleons. This is because the mass of the particle enters the denominator of the expression for the Fermi energy, see Eq. (5.10):

$$E_F = \frac{p_F^2}{2m} = \left(\frac{3}{8\pi}\right)^{2/3}\frac{h^2}{2m}\left(\frac{N}{V}\right)^{2/3}.$$

We could therefore have a situation where

$kT \ll E_F$ (electrons) $\Rightarrow$ degenerate electrons

but $kT \approx E_F$ (nucleons) $\Rightarrow$ nondegenerate nucleons.

So the bottom line is the following. Only the electrons are completely degenerate in a white dwarf. *It is the pressure of the electrons that balances gravity in a white dwarf.* The protons and neutrons are silent spectators. The nominal pressure they exert is only by virtue of their modest thermal motion at the temperature of the white dwarf. This pressure is negligible compared to the pressure exerted by the electrons.

All Stars will Ultimately Find Peace

The mass–radius relation (6.13) predicts finite equilibrium configurations for all stars. *All stars will therefore find peace ultimately as white dwarfs.* Eddington would have found it comforting to know, 'all stars will have the necessary energy to cool'.

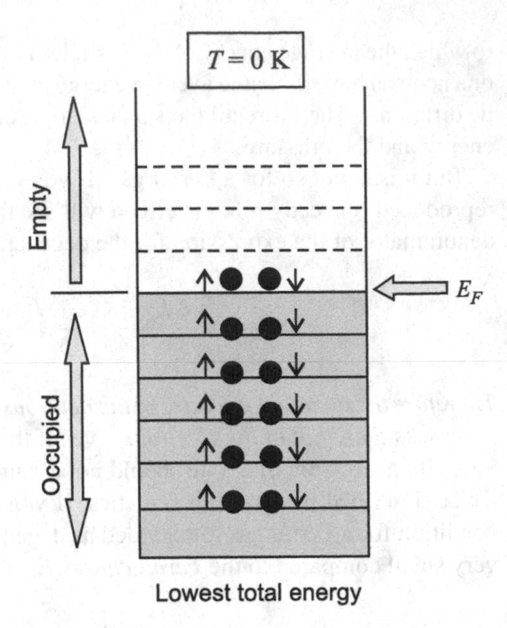

Fig. 6.5 The **ground state** of a degenerate electron gas. *This is the lowest energy configuration*

You might think that there is something fishy in this. After all, gaseous stars were also stable for millions to billions of years. But they got into trouble when their supply of heat was exhausted. How can we be sure that a similar fate is not in store for white dwarfs? What if the internal energy of the white dwarf is radiated away? If that happens, the white dwarf will also be doomed. *But this cannot and will not happen*! It is like a young person inheriting a huge fortune from a rich uncle, but the money is held by a Trust; he or she cannot touch it until they attain a certain age! Similarly, although a white dwarf has an incredible amount of internal energy, it cannot expend it—it is held in trust **forever**. Fowler put it most eloquently in his historic paper of 1926.

> The black dwarf material is best likened to a single gigantic molecule in its lowest quantum state. On the Fermi–Dirac statistics, its high density can be achieved in one and only one way, in virtue of a correspondingly great energy content. But this energy can no more be expended in radiation than the energy of a normal atom or molecule. The only difference between black dwarf matter and a normal molecule is that the molecule can exist in a free state while the black dwarf matter can only so exist under very high external pressure.

So beautifully put! I hope Fowler's reasoning is clear to you. As we stressed in the previous chapter, the enormous energy possessed by the electron gas is *zero point energy* or *ground state energy*. This is like the energy of an electron in the $n = 1$ level of a hydrogen atom. Although the electron has 13.6 eV of energy, *it cannot spend it because there are no allowed levels with lower energy*. Similarly, *a completely degenerate electron gas cannot lower its energy any further than shown in* Fig. 6.5, *without violating Pauli's Exclusion Principle*, which, of course, it cannot do.
So we can say that white dwarfs are forever!

Chapter 7
The Chandrasekhar Limit

Relativistic Stars

Chandrasekhar gathered his results and wrote up a paper in the beginning of 1930. He was still a student at the Presidency College in Madras. By the time he completed his BSc Honours, he had secured a Government of India Scholarship to study under Fowler in Cambridge. He sailed from Bombay (now Mumbai), on 31 July 1930. During the voyage, he started thinking about physics again. Upon reading the manuscript of the paper he had written he began to wonder whether his theory provided a good description for white dwarfs of *all* masses. And the reason for this second thought was the following.

You will recall that the Fermi momentum increases with increasing density as:

$$p_F = \left(\frac{3}{8\pi}\right)^{\frac{1}{3}} h \left(\frac{N}{V}\right)^{\frac{1}{3}} \propto n^{\frac{1}{3}}. \tag{7.1}$$

According to Chandrasekhar's theory, the mean density of a white dwarf increases as the square of the mass, and the central density is six times the mean density (see Fig. 6.4). Chandrasekhar estimated that even in a white dwarf of one solar mass, the central density is so large that the Fermi momentum would be comparable to mc,

$$p_F \sim mc.$$

To put it differently, the electrons on the surface of the Fermi sphere (see Fig. 7.1) will have speeds close to that of light. This meant that the *variation of mass with velocity* predicted by Einstein's *Special Theory of Relativity* must be taken into account in obtaining the equation of state (that is, the expression for the pressure in terms of the density). Obviously, this effect would be much more pronounced in a white dwarf with mass greater than one solar mass. Chandrasekhar decided to obtain the equation of state for a *fully relativistic* electron gas in which *all* the particles, and not just those near the centre of the star, had speeds close to the speed of light, that is, $p \approx mc$

G. Srinivasan, *Life and Death of the Stars*, Undergraduate Lecture Notes in Physics,
DOI: 10.1007/978-3-642-45384-7_7, © Springer-Verlag Berlin Heidelberg 2014

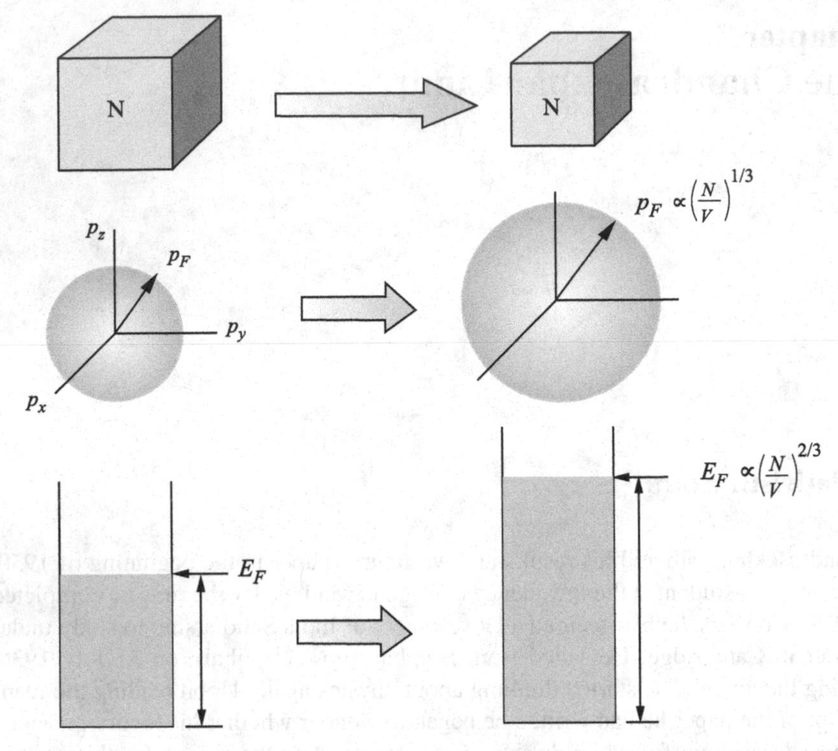

Fig. 7.1 As may be seen in this reproduction of Fig. 5.6, as the density of fermions increases, the maximum momentum, p_F, increases as $n^{\frac{1}{3}}$. The maximum energy, E_F, increases as $n^{\frac{2}{3}}$. When the density becomes sufficiently large, like it does near the centre of a white dwarf, the maximum momentum becomes comparable with mc. Consequently, the variation of mass with velocity predicted by Einstein's special theory of relativity will have to be taken into account while determining the degeneracy pressure

for all particles. He then went on to recalculate the mass–radius relation. Before telling you what he found, let us refresh our memory about the variation of mass with velocity predicted by the special theory of relativity.

Relation Between Mass and Energy in Special Relativity

Let m_o be the mass of a particle at rest; this is known as the *rest mass*. In Newtonian mechanics, the *inertial mass* of the particle (the proportionality constant between force and acceleration in the equation $F = ma$) is just m_o; the mass of the particle is independent of velocity. But according to Einstein, the inertial mass m varies with velocity as:

$$m = \frac{m_o}{\left(1 - v^2/c^2\right)^{\frac{1}{2}}}. \tag{7.2}$$

The momentum p of a particle moving with a velocity v is then given by

$$p = mv = \frac{m_o v}{\left(1 - v^2/c^2\right)^{\frac{1}{2}}}. \tag{7.3}$$

I am sure you know the famous relation between mass and energy,

$$E = mc^2. \tag{7.4}$$

This energy consists of two parts: the energy equivalent of the rest mass, also called the *rest energy* $m_o c^2$ and the *kinetic energy* T_{kin}. We can write

$$T_{kin} = (m - m_o)c^2 = m_o c^2 \left(\frac{1}{\sqrt{1 - v^2/c^2}} - 1 \right). \tag{7.5}$$

Let us now write the total energy of the particle in a form that is useful.

$$E = mc^2 = \frac{m_o c^2}{\left(1 - v^2/c^2\right)^{1/2}}. \tag{7.6}$$

The inertial mass m and the momentum p can be written as

$$m = \frac{E}{c^2} \qquad p = mv = \frac{Ev}{c^2}. \tag{7.7}$$

Squaring both sides of Eq. (7.6) and using Eq. (7.7) we get

$$E^2 = m_0^2 c^4 + p^2 c^2 \tag{7.8}$$

It would be a good exercise for you to convince yourself that when the velocity of the particle is much less than the speed of light ($v/c \ll 1$), Eq. (7.8) will reduce to the Newtonian expression, $E = \frac{1}{2}mv^2$. The relations given above suggest that *inertial mass* may be a property of *energy* rather than of matter as such, each erg of energy possessing, or having associated with it, $1/c^2$ gram of mass. The law of conservation of mass then becomes merely another aspect of the conservation of energy.

Degeneracy Pressure of an Ultra-Relativistic Gas

Let us now derive the relation between pressure and density of an ultra-relativistic electron gas. In the ultra-relativistic limit the kinetic energy of the electron is much greater than its rest mass energy. Therefore, Eq. (7.8) can be approximated as

$$E \approx pc. \tag{7.9}$$

Strictly speaking, Eq. (7.9) is valid only when $v = c$. For a photon gas, $p = h\nu/c$ and Eq. (7.9) reduces to the familiar $E = h\nu$.

We will follow the same steps as in Chap. 5, 'Fermi–Dirac Distribution', with the difference that for an ultra-relativistic gas, $E = pc$ and *not* the Newtonian expression $p^2/2m$; the rest of the steps will be identical. The first step in calculating the pressure is to calculate the total energy of the gas at absolute zero. We shall use the general expression (5.13) for the total energy:

$$E_{\text{Total}} = \int_0^\infty Ef(E)\, g(E)dE.$$

At absolute zero, the probability function $f(E)$ is unity for $E < E_F$ and zero for $E > E_F$ (see Fig. 5.4). Replacing the upper limit of the integration by E_F, we get

$$E_{\text{Total}} = \int_0^{E_F} E\ g(E)\, dE. \tag{7.10}$$

In the above expression, $g(E)dE$ is the number of states in the energy interval being considered. Earlier, we got this from the density of states in momentum space $g(p)dp$ as in Eq. (5.4):

$$g(p)dp = \frac{8\pi V}{h^3} p^2 dp,$$

by using the *nonrelativistic* relation between the momentum and energy, namely, $E = p^2/2m$. This gave us

$$g(E)dE = \frac{8\pi V}{h^3}\sqrt{2m^3}\sqrt{E}dE.$$

In the extreme relativistic limit $E \neq p^2/2m$, but $E = pc$. Using this we obtain

$$\boxed{g(E)dE = \frac{8\pi V}{c^3 h^3} E^2 dE.} \tag{7.11}$$

This expression for the density of states differs from the nonrelativistic expression in two respects.

1. $g(E) \propto E^2$, instead of $\sqrt{E}$.
2. The mass of the particle does not enter the expression.

Using Eqs. (7.11) in (7.10) and integrating, we obtain for the ground state energy of a relativistic electron gas:

$$E_{\text{Total}} = \frac{2\pi V}{c^3 h^3} E_F^4. \tag{7.12}$$

Since $E = pc$ for a relativistic gas, $E_F = p_F c$ (not $p_F^2/2m$). Using the expression (5.8) for the Fermi momentum, we obtain

$$\boxed{E_F = p_F c = \left(\frac{3}{8\pi}\right)^{1/3} hc \left(\frac{N}{V}\right)^{1/3}} \tag{7.13}$$

Notice two important differences between this expression for the Fermi energy and the earlier one in Eq. (5.10). For a relativistic gas,

1. $E_F \propto n^{\frac{1}{3}}$ and not $n^{\frac{2}{3}}$, and
2. *The mass of the particle does not enter the expression.* This is as it should be. In the extreme relativistic case, the rest mass energy of the particle is insignificant compared to the kinetic energy. To put it differently, the energy depends only on the momentum and not on the rest mass.

Note the interesting fact that expression (7.1) for the Fermi momentum remains the same in the nonrelativistic and relativistic case. The Fermi momentum, which has the significance of the maximum momentum of the particles is determined only by the total number of particles and the size of the cells in phase space (refer to Fig. 5.5). Neither of these depends on the speed of the particle. As you will see in Chap. 8, 'The Absurd Behaviour of Stars', although correct, this conclusion got Chandrasekhar into a great deal of trouble!

Substituting Eq. (7.13) in (7.12), we obtain for the total energy at absolute zero,

$$\boxed{E_{\text{Total}} = V\frac{3}{4}\left(\frac{3}{8\pi}\right)^{\frac{1}{3}} hc \left(\frac{N}{V}\right)^{\frac{4}{3}}.} \tag{7.14}$$

The pressure of a *relativistic gas* is related to the internal energy by

$$P = \frac{1}{3}\frac{E_{\text{Total}}}{V}. \tag{7.15}$$

In the nonrelativistic case described by Eq. (5.19), the prefactor was 2/3. There are many ways of understanding Eq. (7.15) but we shall not get into it here. If you recall, radiation pressure is also one-third of the energy density of radiation in the cavity:

$$p_{\text{rad}} = \frac{1}{3} a T^4$$

The general rule is the following. If one is dealing with particles with finite rest mass, the prefactor relating pressure and energy density is 2/3. For radiation, as well as particles which may be assumed to travel at speeds close to that of light, the prefactor is $1/3$. Combining Eqs. (7.14) and (7.15) we finally obtain for the pressure of a relativistic electron gas:

$$\boxed{P_{\text{rel}} = \frac{1}{3} \frac{E_{\text{Total}}}{V} = \frac{1}{8} \left(\frac{3}{\pi}\right)^{\frac{1}{3}} hcn^{\frac{4}{3}}.} \tag{7.16}$$

Written in terms of the mass density:

$$\boxed{P_{\text{rel}} = \mathrm{K}_2 \rho^{\frac{4}{3}}} \tag{7.17}$$

The constant K_2 is given by

$$\mathrm{K}_2 = \frac{1}{8} \left(\frac{3}{\pi}\right)^{\frac{1}{3}} \frac{hc}{(\mu_e m_p)^{\frac{4}{3}}}. \tag{7.18}$$

Let us summarize.

1. *The pressure of a relativistic gas is proportional to* $\rho^{4/3}$. In the nonrelativistic case, it is proportional to $\rho^{5/3}$.
2. *The pressure of a relativistic gas is independent of the mass of the particle, unlike in the nonrelativistic case.*

The above results were derived for the first time by Chandrasekhar during his voyage to England.

A Startling Discovery by Chandrasekhar

Having obtained the equation of state of a relativistic gas, Chandrasekhar took the next step to model such completely relativistic stars. The procedure was the same as discussed in Chap. 6, 'Quantum Stars'. One has to solve the equation of hydrostatic equilibrium:

$$\frac{dP}{dr} = -\frac{GM(r)\rho(r)}{r^2}, \tag{7.19}$$

with the pressure now being given by equation (7.17):

$$P_{\rm rel} = K_2 \rho^{\frac{4}{3}}.$$

This is again a polytropic equation of state:

$$P = K\rho^{1+\frac{1}{n}}$$

with $n = 3$. Chandrasekhar knew how to solve this problem in a mathematically exact manner. He knew that in the case of a polytrope of index $n = 3$, the radius of the star is uniquely determined by K_2, the constant of proportionality in the pressure–density relation (7.17). Before telling you the exact result he found, let us, as we did in the nonrelativistic case, do an elementary derivation and prepare ourselves for the shock! Let us, once again, make the approximation

$$\frac{dP}{dr} \approx \frac{(P(r) - P_{\rm surface})}{R} = \frac{P}{R}.$$

With this approximation equation (7.19) for equilibrium gives

$$\frac{P}{R} \propto \frac{M\rho}{R^2}.$$

Let us now use $P \propto \rho^{\frac{4}{3}}$ and $\rho \propto \frac{M}{R^3}$ in the above equation. We get

$$\frac{1}{R}\left(\frac{M}{R^3}\right)^{\frac{4}{3}} \propto \frac{M}{R^2}\left(\frac{M}{R^3}\right).$$

Simplifying, we get

$$\boxed{\frac{M^{\frac{4}{3}}}{R^5} \propto \frac{M^2}{R^5}}. \qquad (7.20)$$

This is a most surprising and extraordinary result. In the nonrelativistic case, the same steps led us to a relation between mass and radius of the star (refer to Eqs. 6.14 and 6.15). Equation (7.20) does not, however, yield a relation between the mass and the radius. *The radius, in fact drops out of the result since the same power of the radius appears on both sides of the above equation!* The only variable that remains is the mass.

We are thus forced to conclude from Eq. (7.20) that

1. *A completely relativistic star has no radius,*
2. However, *a completely relativistic star has a unique mass.*

Our simple scaling argument cannot give the value of this unique mass. Since Chandrasekhar solved the equation for equilibrium exactly, he was able to derive its value. He obtained

$$M = 4\pi \left(\frac{K_2}{\pi G}\right)^{\frac{3}{2}} 6.89.$$

Substituting for K_2 from Eq. (7.18) and simplifying

$$\boxed{M_{\mathrm{Ch}} = 0.197 \left[\left(\frac{hc}{G}\right)^{\frac{3}{2}} \frac{1}{m_p^2}\right] \times \frac{1}{\mu_e^2}} \tag{7.21}$$

This is one of the most beautiful results in physics, and let us savour it. In ***What Are the Stars?*** we asked the question, 'Why do the stars have nearly the same mass?' The lightest known star is about 3×10^{32} g, and the heaviest about 2×10^{35} g. The majority are between 10^{33} and 10^{34} g. We argued that Eddington's theory of the stars in which gravity is balanced by the combined pressure of the gas and radiation isolates a combination of fundamental constants with the dimension of mass, and that this gives *a characteristic scale for the masses of the stars*. This combination of fundamental constants was

$$\boxed{\left(\frac{hc}{G}\right)^{\frac{3}{2}} \frac{1}{m_p^2} \cong 29.2 M_{\odot}.}$$

Note that the numerical value of this *mass scale* is 29.2 times the mass of the Sun. Thus, a star which may be described by Eddington's theory must have mass which is a few times the mass of the Sun. His theory tells us that the characteristic mass of stars would be a few times 10^{33} g (recall that the mass of the Sun is 2×10^{33} g). Eddington's stars will not be of planetary mass, nor would they be thousands of times more massive than 10^{33} g.

Once again, we encounter the *same combination of fundamental constants* in Eq. (7.21). This time, however, it does not provide a *scale* for the measurement of masses. *It gives a unique value for the mass of a completely relativistic white dwarf.* Let us go back to the parable of the physicists on a cloud-bound planet that we discussed at length in, ***What Are the Stars?*** This time, let the physicists construct objects in which gravity is balanced by the degeneracy pressure of relativistic electrons. To their great surprise, they would find that such stars would have a unique mass determined solely by a combination of fundamental constants. The numerical value of this unique mass is

$$\begin{aligned} M_{\mathrm{Ch}} &= 0.197 \left[\left(\frac{hc}{G}\right)^{\frac{3}{2}} \frac{1}{m_p^2}\right] \times \frac{1}{\mu_e^2} \\ &= 0.197 \times 29.2 M_{\odot} \times \frac{1}{\mu_e^2} = 5.76 M_{\odot} \times \frac{1}{\mu_e^2}. \end{aligned} \tag{7.22}$$

For an assumed value of $\mu_e = 2$, we get

$$\boxed{M_{\mathrm{Ch}} = 1.4 M_{\odot}} \tag{7.23}$$

This unique mass of a fully relativistic white dwarf given by Eq. (7.21) has come to be known as the *Chandrasekhar Limiting Mass*. It is important to stress that this is an exact result.

Whereas Chandrasekhar did not take much time to discover this startling result, he could not make head or tail of it! Why it was appropriate to call it a *limiting mass* became clear only four years later, in 1934.

The Chandrasekhar Limit

After a long journey, Chandrasekhar reached Cambridge early in September 1930. But he had to wait for a month before his first meeting with Fowler, just a few days before the semester started. The first thing he did was to show the paper on his theory of the white dwarfs. Fowler was very appreciative of it and complimented him. Next, Chandra showed Fowler the intriguing result he had obtained during the voyage. Regarding this, Fowler was sceptical. But he did send it to the famous astronomer E.A. Milne for his learned comments. Unfortunately, Milne was also very sceptical of the result on the limiting mass.

As Chandra was to remark later, he himself was intrigued, puzzled and confused by the result he had obtained. What did it mean? Soon a picture emerged in his mind. Perhaps the relation $R \propto M^{-1/3}$ given by the nonrelativistic theory was modified by the inclusion of the relativistic effects in the following way. Consider a white dwarf as consisting of a nonrelativistic *envelope* (in which $P \propto \rho^{5/3}$) and a relativistic *core* (in which $P \propto \rho^{4/3}$), as shown in Fig. 7.2. As we go to more massive white dwarfs, one would expect that the envelope will shrink in mass and the core will grow in mass. The completely relativistic model, considered as a limit of such composite stars is a point mass with $\rho = \infty$!

This was Chandra's conjecture, but he was still puzzled. If the critical mass he had derived is a *limiting mass*, then what is the fate of stars more massive than $1.4 M_{\odot}$? Even though the true significance of his result eluded him, he was convinced of its correctness and potential importance. He therefore submitted it for publication. Since both Fowler and Milne were lukewarm in their responses, he sent it to *The Astrophysical Journal*, published in America, rather than a British journal! The paper was published in 1931. Although ignored for more than three decades, this paper is now recognized as one of the most significant papers in contemporary astronomy.

Although the question concerning the ultimate fate of the stars was not settled, Chandra had to move on and worry about his thesis. Given the uncertainty, he chose an entirely different topic for his thesis, a study of the interaction between radiation and matter in the atmosphere of stars (a work that brought him much recognition and fame).

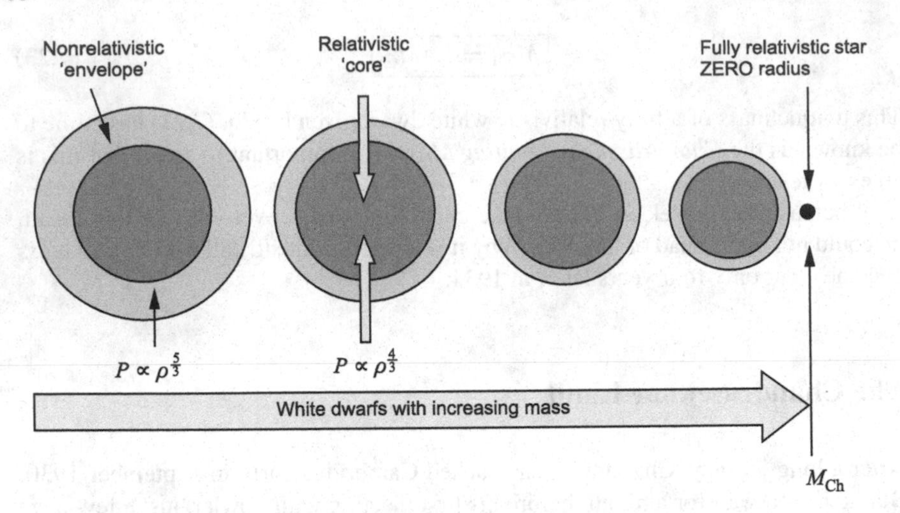

Fig. 7.2 Chandrasekhar conjectured that the unique mass he had obtained for a fully relativistic white dwarf may be visualized as a *limit* of the sequence shown above. In the limit $M \to 0$, all the electrons in the white dwarf will be nonrelativistic, but as one goes to more massive white dwarfs, the electrons near the centre will be relativistic and one may think of them as forming a *core*. As we go to more massive stars, the mass contained in the relativistic core will increase. Note, however, the *size of the core will decrease* because of the inverse relationship between the radius and mass in degenerate stars! Finally, when the entire star becomes relativistic at $M = M_{Ch}$, *the radius goes to zero and the density becomes infinite*

But he returned to white dwarfs in 1934. By that time he had obtained his doctorate and secured the much coveted *Fellowship of Trinity College, Cambridge* (the only other Indian to have secured this most prestigious Fellowship was the legendary mathematician **Srinivasa Ramanujan**). This time he decided to tackle this problem head on, and not make any approximations. Remember that he had calculated the internal constitution of white dwarfs in two limiting cases:

1. When the velocity of the particles was small compared to the velocity of light. In this limit, Newtonian mechanics is valid and the kinetic energy of the particles is given by $E = \frac{1}{2}m\text{v}^2$. The degeneracy pressure of the electrons was given by $P = \text{K}_1\rho^{5/3}$, as shown in Eq. (6.10).
2. When the velocity of the electrons was very nearly equal to the speed of light. In this extreme relativistic limit, $E = pc$, like for the photon gas. The pressure is given by $P = \text{K}_2\rho^{4/3}$, as shown in Eq. (7.17).

In a real white dwarf, not all the electrons will have speed $v \sim c$. While the electrons near the centre (where the density is six times the average density) may be relativistic, those in the outer region may have speed $v \ll c$. Therefore, the correct thing to do would be to use the general expression for the energy as in Eq. (7.8)

$$E = \left(p^2c^2 + m_0^2c^4\right)^{\frac{1}{2}} . \tag{7.24}$$

As remarked earlier, this expression reduces to $E = \frac{1}{2}mv^2$ in the limit $v \ll c$, and $E = pc$ in the limit $v \sim c$. One can derive the expression for the degeneracy pressure by using Eq. (7.24) for the energy of the particle. The derivation will proceed along the same lines as before. Chandrasekhar did this, and the expression for the pressure that he obtained is

$$P = \mathrm{A}f(x), x = \left(\frac{\rho}{\mathrm{B}}\right)^{\frac{1}{3}}, \tag{7.25}$$

where

$$\mathrm{A} = \frac{\pi m^4 c^5}{3h^3}, \qquad \mathrm{B} = \frac{8\pi m^3 c^3 \mu_e m_p}{3h^3} \tag{7.26}$$

and

$$f(x) = x(x^2+1)^{1/2}(2x^2-3) + 3\sinh^{-1} x. \tag{7.27}$$

Although this looks horribly complicated, it is quite straightforward to derive the above expression. If you have the patience you might verify that in the limit of low electron density ($x \ll 1$) the above expression reduces to the nonrelativistic result $P = \mathrm{K}_1\rho^{5/3}$, and in the limit of high electron density ($x \gg 1$) we recover our earlier result in the relativistic limit $P = \mathrm{K}_2\rho^{4/3}$.

The next step was to derive the exact mass–radius relation. Again, we have to repeat our earlier steps, this time with the equation of state given by the expression (7.25). Today, modern computers would make this a simple exercise. But this was quite an involved thing to do in 1934. Undeterred, Chandra plunged in and calculated the mass–radius relation point by point. It was heavy numerical work. At some stage Eddington borrowed a mechanical hand calculator from a Norwegian visitor and gave it to Chandra; this Swedish calculator was the only one around! After many months of labour, Chandra derived the exact mass–radius relation, which is schematically shown in Fig. 7.3.

Let us look at this figure carefully. The dashed curve is the approximate theory which neglects special relativistic effects, namely, the variation of mass with velocity. That theory predicts an equilibrium configuration for all masses. The solid line is the *exact* theory. As would be expected, the exact theory predicts the same mass–radius relation as the approximate theory for $M \to 0$. In white dwarfs having very low masses, the density is small enough that the electrons do not have velocities comparable to light. Naturally, special relativistic effects are unimportant. But the exact theory deviates from the approximate theory as one goes to more massive white dwarfs, takes a nose dive around one solar mass and the radius *tends to zero* for $M \to M_{\mathrm{Ch}}$. Therefore, *finite equilibrium configurations exist only for* $M < M_{\mathrm{Ch}}$.

Let us pause to understand the implications of this result. Eddington was worried that the stars like the companion of Sirius 'did not have enough energy to cool'. Fowler rescued these stars. He argued that stars will eventually pass into a white dwarf stage and cool, thus becoming black dwarfs. But Chandrasekhar's remarkable discovery showed that *stars more massive than* M_{Ch} *cannot find equilibrium as white dwarfs*,

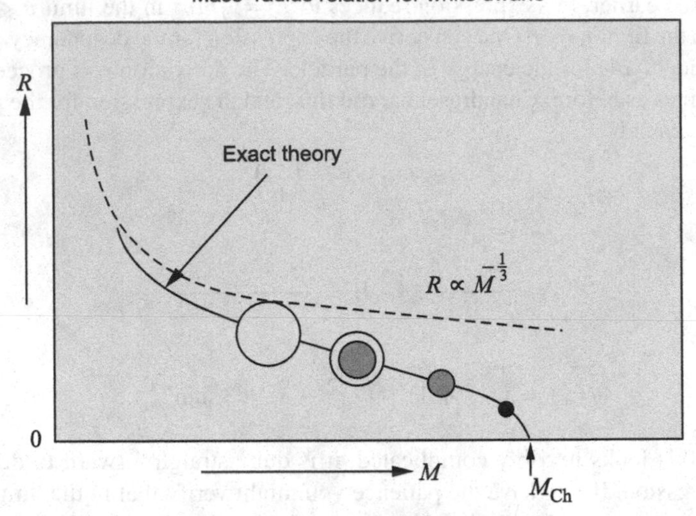

Fig. 7.3 This figure has been adapted from the historic paper by Chandrasekhar (1935) in which he presented his exact theory of white dwarfs. The *dashed curve* is the approximate theory in which the electrons were regarded as nonrelativistic. The solid curve is the exact mass–radius relation. We see that that the approximate theory agrees with the exact theory in the limit of very low mass. But the exact theory deviates from the approximate theory as one goes to more massive white dwarfs, takes a nose dive around one solar mass, and *the radius tends to zero for* $M \to M_{\rm Ch}$. For an explanation of the circles superimposed on the exact curve, refer to Fig. 7.2

Hence, *they will not have the necessary energy to cool.* Thus, stars with $M > M_{\rm Ch}$ *will* be in an awkward predicament when their supply of subatomic energy fails!

Can All Stars Find Peace?

The exact theory of Chandrasekhar confirmed his earlier intuition that the unique mass he had discovered:

$$M_{\rm Ch} = 0.197 \left[\left(\frac{hc}{G} \right)^{\frac{3}{2}} \frac{1}{m_p^2} \right] \times \frac{1}{\mu_e^2} = 1.4 M_{\odot}.$$

should be interpreted as a *limiting mass* of white dwarfs. What then is the fate of stars more massive than this? Interestingly, Chandrasekhar had found the answer to this question in 1932 itself, even before he had found an exact solution of the white dwarf problem. In a remarkable paper published in 1932, he had obtained the fundamental result:

If radiation pressure is more than 9.2 percent of the total pressure (gas pressure plus radiation pressure)then matter cannot become degenerate, however high the density may become.

We shall now prove this.

You will remember from our discussion of the Fermi–Dirac distribution that fermions should be regarded as degenerate if $E_F \gg kT$, that is, if the Fermi energy is much greater than the thermal energy. Alternatively, one can say that a gas should be considered as degenerate if the numerical value of the pressure calculated from the Fermi–Dirac distribution is much greater than the value of the pressure calculated using Boyle's law,

$$p_{\mathrm{deg}} \gg p_{\mathrm{ideal\ gas}}. \tag{7.28}$$

Conversely, *if the classical pressure exerted by the electrons is much greater than the degeneracy pressure of the electrons then matter cannot be regarded as degenerate*. You might think there is something fishy in this. After all, Fowler came to the conclusion that the matter inside the companion of Sirius should be regarded as degenerate because $E_F \gg kT$. This implies that $p_{\mathrm{deg}} \gg p_{\mathrm{ideal\ gas}}$. You can easily convince yourself of this as follows. The nonrelativistic degeneracy pressure of the electrons is given by

$$p_{\mathrm{deg}} = \frac{2}{5}\left(\frac{3}{8\pi}\right)\frac{h^2}{2m}\left(\frac{N}{V}\right)^{\frac{5}{3}} = \frac{2}{5}\left(\frac{N}{V}\right)E_F. \tag{7.29}$$

(refer to Eqs. (5.10) and (5.20)), while the classical pressure of the electrons is given by

$$p_e = \left(\frac{N}{V}\right)kT. \tag{7.30}$$

A comparison of Eqs. (7.29) and (7.30) shows that since $E_F \gg kT$ for a degenerate gas, it follows that $p_{\mathrm{deg}} \gg p_{\mathrm{ideal\ gas}}$.

Why should Chandrasekhar entertain the possibility that this condition may be reversed in more massive stars? After all, one would expect the density to be even greater in a more massive star. But then, the internal temperature would also be greater in a more massive star. Since the classical pressure given by Eq. (7.30) depends upon *both the density and temperature*, one cannot assume that the inequality (7.28) would necessarily be satisfied in a star more massive than the Chandrasekhar limiting mass. That is why Chandrasekhar investigated the general condition for matter to be regarded as degenerate.

However, comparing degeneracy pressure and classical pressure is like comparing apples and oranges! While the former depends only on the density, the latter depends on both the density and temperature. But there is a little trick we employed in ***What Are the Stars?***, which we shall use once again. Let us introduce a fraction β defined as follows:

$$\begin{aligned} P_{\text{tot}} &= p_e + p_{\text{rad}} \\ &= \frac{1}{\beta} p_e = \frac{1}{1-\beta} p_{\text{rad}} \\ &= \frac{1}{\beta} \frac{\rho k T}{\mu_e m_p} = \frac{1}{1-\beta} \frac{1}{3} a T^4 \end{aligned} \tag{7.31}$$

The meaning of β is clear. *It is the fraction of the total pressure contributed by the classical pressure p_e of the electrons, while $(1-\beta)$ is the fraction due to the radiation pressure p_{rad}*. Let us equate the two sides in the last of the three equations above and express T in terms of β and ρ:

$$\frac{1}{\beta} \frac{\rho k T}{\mu_e m_p} = \frac{1}{1-\beta} \frac{1}{3} a T^4.$$

Simplifying, we obtain

$$\boxed{T = \left[\frac{3}{a} \frac{k}{\mu_e m_p} \frac{1-\beta}{\beta} \right]^{\frac{1}{3}} \rho^{\frac{1}{3}}.} \tag{7.32}$$

We can now use Eq. (7.32) to substitute for T in the expression for the classical pressure of the electrons:

$$\boxed{p_e = \frac{\rho k T}{\mu_e m_p} = \left[\frac{3}{a} \left(\frac{k}{\mu_e m_p} \right)^4 \left(\frac{1-\beta}{\beta} \right) \right]^{\frac{1}{3}} \rho^{\frac{4}{3}}.} \tag{7.33}$$

We have succeeded in writing the pressure of an ideal gas in terms of ρ and β, instead ρ and T. We can now compare apples and oranges. But what do we compare it with—the nonrelativistic degeneracy pressure or the relativistic version? Chandrasekhar's exact theory of white dwarfs established beyond doubt that the electrons are fully relativistic by the time we reach a mass of $1.4 M_\odot$ in the sequence of white dwarfs. It is therefore reasonable to suppose that if degeneracy sets in at all in stars more massive than $1.4 M_\odot$, it will do so with the electron fully relativistic. Therefore, *we should compare the ideal gas pressure of the electrons with the relativistic degeneracy pressure $P_{\text{rel}} = K_2 \rho^{\frac{4}{3}}$* (see Eqs. (7.17) and (7.18)).

Again we have a dilemma. If electrons are relativistic, are we justified in using Boyle's law to calculate the ideal gas pressure of the electrons? After all, Boyle's law predates special relativity by several centuries. Yes, *Boyle's law is valid in relativity also*! To quote Chandrasekhar from his 1932 paper,

> In this connection it will have to be remembered that considerations of relativity do not affect the equation of state of a perfect gas. $p = nkT$ is true independent of relativity!

One cannot but admire this clarity of thinking by a twenty-two year old boy! Therefore *the classical pressure of the electrons will be greater than the degeneracy pressure*, that is, $p_e > P_{\rm rel}$, if

$$p_e = \frac{\rho k T}{\mu_e m_p} = \left[\frac{3}{a}\left(\frac{k}{\mu_e m_p}\right)^4\left(\frac{1-\beta}{\beta}\right)\right]^{\frac{1}{3}} \rho^{\frac{4}{3}} > \mathrm{K}_2 \rho^{\frac{4}{3}} = P_{\rm rel}.$$

Cancelling the density dependence on both sides, we obtain the condition for $p_e > P_{\rm rel}$ as

$$\left[\frac{3}{a}\left(\frac{k}{\mu_e m_p}\right)^4\left(\frac{1-\beta}{\beta}\right)\right]^{\frac{1}{3}} > \mathrm{K}_2. \tag{7.34}$$

Substituting for K_2 from Eq. (7.18), and inserting the value for Stefan's constant, a, the above inequality reduces to

$$\frac{960}{\pi^4}\frac{1-\beta}{\beta} > 1.$$

This can be recast as

$$\boxed{1 - \beta > 0.092.}$$

Recall that $(1-\beta)$ is the fraction of the total pressure due to radiation,

$$p_{\rm rad} = (1-\beta) P_{\rm tot}.$$

Thus, *if radiation pressure is greater than 9.2 percent of the total pressure, then $p_e > P_{\rm rel}$, and matter cannot be regarded as degenerate.* As we shall see in later chapters, this exact result is of singular importance in all the contemporary schemes of stellar evolution. This conclusion has been schematically represented in Fig. 7.4.

You may remember from Volume 1, ***What Are the Stars?***, that one of Eddington's important insights was that the magnitude of the radiation pressure increases with increasing mass. In the centre of the Sun, for example, radiation pressure is only about 3 percent of the total pressure. As one goes to more massive stars, one would encounter a *critical mass,* $M_{\rm critical}$ in which radiation pressure is precisely equal to 9.2 percent of the total pressure. Therefore, matter cannot become degenerate in stars more massive than this critical mass. Consequently, eventually attaining a white dwarf stage is impossible for these stars. These stars will be in serious trouble when their supply of nuclear energy is exhausted. Since Pauli's exclusion principle and Fermi–Dirac statistics cannot save them, they will have no option but to collapse indefinitely until they become point objects with infinite density!

Using Eddington's *Standard Model* for the stars, Chandrasekhar estimated this critical mass to be approximately $1.6 M_\odot$, just a little more than the limiting mass for white dwarfs. But this estimate of the critical mass based on the Standard Model

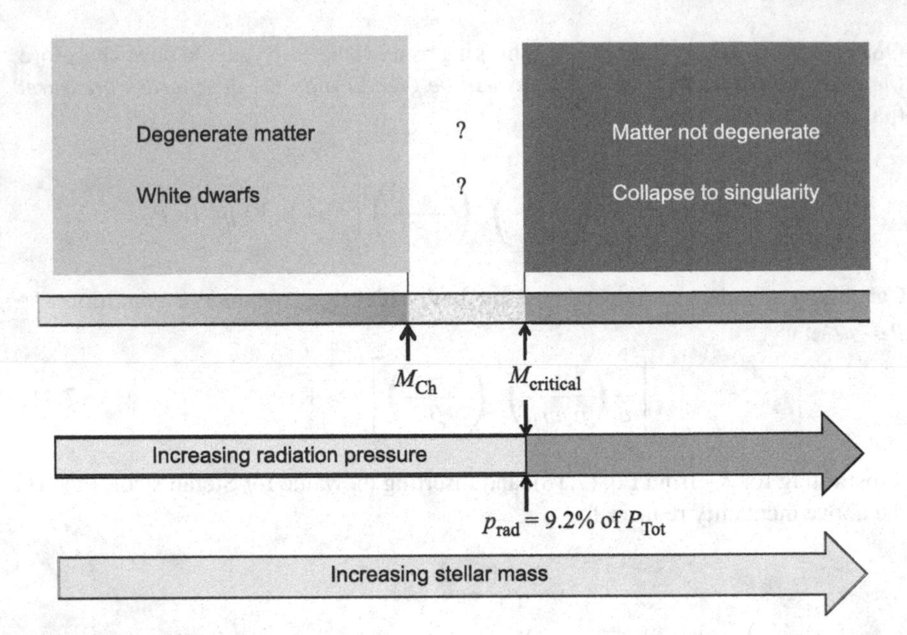

Fig. 7.4 This figure schematically summarizes two startling discoveries made by Chandrasekhar. (1) There is a limiting mass for white dwarfs. This limiting mass, known as the *Chandrasekhar limit*, M_{Ch}, is determined solely by fundamental constants. Its numerical value is $1.4M_{\odot}$ (on the assumption that the star had exhausted all the hydrogen before becoming degenerate). (2) There is a *critical mass* for stars. Stars with mass greater than this critical value will not become degenerate, however high the density may become. Such stars cannot be saved by Pauli's exclusion principle when their supply of nuclear energy is exhausted. They will have no option but to collapse to a singularity

is not correct. However, as we shall see later in this volume, *modern considerations do show that radiation pressure plays an increasingly dominant role as the mass of the star increases, and that stars with mass greater than about* $8M_{\odot}$ *do not develop degeneracy in their interior.* Thus the basic arguments due to Chandrasekhar of nearly eighty years ago, summarized in Fig. 7.4, have been sustained!

Chapter 8
The Absurd Behaviour of Stars: Not All Stars will have Energy to Cool

Some Remarkable Assertions

As we saw, Chandrasekhar made two remarkable discoveries during the short period 1930–1934.

1. In 1930, he stumbled upon the intriguing result that a completely relativistic degenerate star has a unique mass:

$$M_{Ch} = 0.197 \left[\left(\frac{hc}{G} \right)^{\frac{3}{2}} \frac{1}{m_p^2} \right] \times \frac{1}{\mu_e^2} = 1.4 M_{\odot}.$$

   Chandrasekhar established the true significance of the above result through detailed numerical calculations done in 1934. His exact theory of white dwarfs clearly showed that the above mass should be regarded as a limiting mass of white dwarf configurations. Stars more massive than this limiting mass cannot find equilibrium as white dwarfs.
2. In 1932, Chandrasekhar proved a theorem that showed that matter cannot become degenerate if radiation pressure exceeds 9.2 percent of the total pressure. It followed from this theorem that sufficiently massive stars can never develop degenerate cores during their evolution.

These two discoveries are now seen to be at the base of the present revolution in astronomy. Chandrasekhar was so confident of the correctness of his results that his epochmaking papers contained some bold and emphatic statements.

For example, his 1932 paper contained the following statements:

> For all stars of mass greater than $M_{critical}$ the perfect gas equation of state does not break down, however high the density may become, and the matter does not become degenerate. An appeal to the Fermi–Dirac statistics to avoid the central singularity cannot be made.

G. Srinivasan, *Life and Death of the Stars*, Undergraduate Lecture Notes in Physics,
DOI: 10.1007/978-3-642-45384-7_8,

> Great progress in the analysis of stellar structure is not possible before we can answer the following fundamental question: Given an enclosure containing electrons and atomic nuclei (total charge zero) what happens if we go on compressing the material indefinitely?
>
> S. Chandrasekhar (1932)

In a preliminary announcement of his exact result, he concluded:

> Finally, it is necessary to emphasize one major result of the whole investigation, namely, that it must be taken as well established that the life history of a star of small mass must be essentially different from the life history of a star of large mass. For a star of small mass the natural white dwarf stage is an initial step towards complete extinction. A star of large mass ($>M_{critical}$) cannot pass into the white dwarf stage, and one is left speculating on other possibilities.
>
> S. Chandrasekhar (1934)

Eddington's Tirade

These were remarkably prescient statements. And they have stood the test of time. But at that time, the entire astronomical community chose to ignore the remarkable discoveries made by young Chandra. Their attitude of benign neglect was prompted by the fact that some of the High Priests of astronomy had openly declared that Chandra had got it all wrong. I have already remarked that both Fowler and Milne were very sceptical about the result on the completely relativistic degenerate star that Chandra had obtained during his voyage to England. Fowler did not understand it, just as Chandra did not understand it at that time. Milne did not accept Chandra's discovery because it contradicted his theory according to which *all* gaseous stars had degenerate cores. So the notion that degenerate cores cannot exist in stars with mass exceeding a limit was unacceptable to him. Faced with this reaction, and anticipating difficulties in getting his paper published in a British journal, Chandra sent his 1932 paper to a famous German journal for publication. But as luck would have it, the journal sent his paper to Milne for his critical advice. Although Milne (reluctantly) approved the paper for publication, he wrote a letter to Chandra in which he said, '... the paper contains a mistake in principle, and in any case it would only do your reputation harm if it were printed'.

As for Eddington, he was confident that an exact theory of degenerate stars would show that there is no such thing as a limiting mass. So why panic at this stage? As already mentioned, Chandra returned to this problem after completing his Doctoral Thesis, and after securing the prestigious Fellowship of Trinity College. Chandra went to Russia in July 1934. One of the places he visited was the famous Pulkovo Observatory in Leningrad. There he met the renowned Armenian astronomer Victor Ambartsumian, who encouraged him to work out the theory of white dwarfs without making any approximations. He embarked on this soon after he returned to Cambridge. Eddington took great interest in the progress of Chandra's calculations. He must have been aghast to see the radius of the white dwarf taking a nose dive to zero as the mass approached the limiting mass, but he did not express any view and played his cards close to his chest.

In January 1935, Chandra was invited to present a paper at the meeting of the *Royal Astronomical Society* in London. It was in this historic meeting that Chandra presented the conclusions of his exact theory of white dwarfs. Soon after his presentation, the President of the Society invited Eddington to speak. Chandra knew that Eddington was going to speak but he was not aware of what the theme of his talk would be. He was therefore aghast when Eddington rose to speak on the topic, *relativistic degeneracy.* Naturally, this came as a great shock to Chandra. But worse was to follow. The main thrust of Eddington's talk was that there is no such thing as relativistic degeneracy and, therefore, Chandrasekhar's conclusions must be summarily rejected. He began by discussing the history of the problem, the paradox he had posed in 1924, and the way Fowler had solved the problem by invoking Fermi–Dirac statistics. Eddington was annoyed that Chandrasekhar had resurrected the original paradox. Let me quote a few sentences from Eddington's speech:

> I do not know whether I shall escape from this meeting alive, but the point of my paper is that there is no such thing as relativistic degeneracy!
>
> Chandrasekhar, using the relativistic formula which has been accepted for the last five years, shows that a star of mass greater than a certain limit M remains a perfect gas and can never cool down. The star has to go on radiating and radiating, and contracting and contracting until, I suppose, it gets down to a few km radius, when gravity becomes strong enough to hold in the radiation, and the star can at last find peace.
>
> ... Dr. Chandrasekhar had got this result before, but he has rubbed it in in his last paper; and, when discussing with him, I felt driven to the conclusion that this was almost a *reductio ad absurdum* of the relativistic degeneracy formula. Various accidents may intervene to save the star, but I want more protection than that. *I think there should be a law of Nature to prevent a star from behaving in this absurd way.*

He then went on to discuss where he thought the idea of relativistic degeneracy had gone wrong. His contention was that the formula for relativistic degeneracy is based on a combination of relativistic mechanics and nonrelativistic quantum theory. The implication of his remarks was that Pauli's exclusion principle may not be valid in relativistic quantum mechanics! Eddington claimed that when quantum statistical mechanics is properly formulated within the framework of relativity, the *old* formula, $P = K_1\rho^{5/3}$, will, once again, prevail, and that Fowler's solution of the 1924 paradox will hold good for all stars. According to Eddington, stars will not behave in the absurd manner indicated by Chandrasekhar's theory. *All stars will have energy to cool!*

Well, the High Priest had spoken, and the bandwagon started rolling. Many astronomers, notably Milne, climbed on the bandwagon to express their objection to Chandrasekhar's discovery. Needless to say, young Chandra was shattered. Instead of being hailed as a rising superstar in science, he was ridiculed. Faced with this situation Chandra did the only thing he could to counter the mighty Eddington. He appealed to the physicists at Neils Bohr's Institute in Copenhagen. Since he had spent one year there (in 1932) he knew many of them personally. During that period, Neils Bohr's Institute was the 'Mecca of Physics'. All the young geniuses were attracted to Copenhagen. Bohr had a great reputation as a teacher, philosopher and as someone who inspired people to do great things. And this list was truly impressive: Fowler, Dirac, Heisenberg, Pauli, Jordon, Max Born, Oskar Klein, Leon Rosenfeld, Victor Weisskopf, Max Delbrück, and many more! In one of his letters to his father Chandra wrote: 'It could be said only of Bohr that he is not only a great mind but one whose influence on the contemporary geniuses has been colossal. In fact in the whole range of mathematics and physics history, it would be difficult to find Bohr's equal—at the moment I can think of only one name—Gauss'.

So Chandra wrote to his close friend Leon Rosenfeld, explaining Eddington's objections to relativistic degeneracy, in particular to Pauli's Principle in relativity. Rosenfeld discussed Eddington's objections with Bohr and wrote back to Chandra:

> Bohr and I are absolutely unable to see any meaning in Eddington's statements ... It seems to us that Eddington's statement that several high-speed electrons might be in one cell of the phase space would imply that to another observer several slow speed electrons, in contrast to Pauli's Principle, would be in the same cell ... Could you perhaps induce Eddington to state his views in terms intelligible to humble mortals? ...

(Rosenfeld had simply demolished Eddington's argument! Let us understand Rosenfeld's counter attack. The only way Eddington could reject Chandrasekhar's discovery was to argue that Pauli's Exclusion Principle was not valid in Special Theory of Relativity. He agreed with Fermi and Dirac that as long as the electrons were *slow* (that is, nonrelativistic), one can put only two electrons in each cell in phase space. But he maintained that if the electrons were *fast* (or relativistic) one could put as many electrons in each cell in phase space as one wanted. The whole point about Special Theory of Relativity is that what appears a fast electron to one observer might be a slow electron for another observer. Eddington might think he has packed a cell with *fast* electrons. But to another observer, the cell is packed with *slow* electrons, which *would* be a violation of Pauli's Principle! This trivial mistake by Eddington is astonishing because he had written an authoritative book on Einstein's Theory of Relativity).

A few days later Chandrasekhar sent a copy of Eddington's manuscript to Rosenfeld requesting him to show it to Pauli and Bohr. Rosenfeld replied:

> ... After having courageously read Eddington's paper twice, I have nothing to change my previous statements; it is the wildest nonsense!

Pauli's reaction was characteristic: 'Eddington did not understand physics'. Chandra also wrote to Dirac. He, too, thought that there was absolutely nothing wrong with Chandra's treatment of the problem.

Although the great physicists were convinced that Eddington's objections were ridiculous, they did not want to openly confront him. They simply could not be bothered. At that time, physicists were not interested in astrophysical problems. Remember that at that time the problem of energy generation in the Sun had not yet been solved. Bohr, for example, felt that since astronomers could not answer even that basic question, it was premature for physicists to get involved in astronomy. Interestingly, the problem of energy generation was eventually solved in 1938 by the *physicist* Hans Bethe! Another brilliant physicist who was into astrophysical

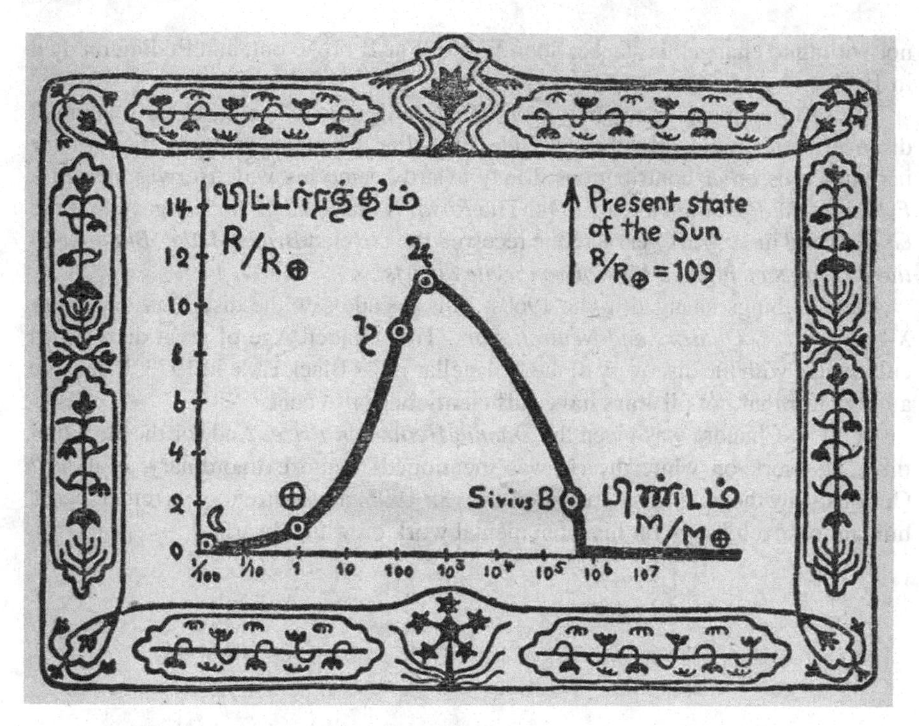

Fig. 8.1 This figure has been reproduced from George Gamow's book, *The Birth and Death of the Sun*. The symbols on the rising potion of the curve represent (from *left* to *right*) the Moon, the Earth, Saturn and Jupiter. To quote Gamow: 'Note that for masses greater than 460,000 times the mass of the Earth the radius becomes zero! *The words for mass and radius are in Dr Chandrasekhar's original Tamil*'

problems even at that time was *George Gamow* (in fact, it was Gamow who got Bethe interested in the stellar energy problem). Gamow was not only a most creative physicist; he was also a great communicator and wrote many wonderful popular books explaining the most recent developments in physics. In one of them, entitled, *The Birth and Death of the Sun,* and published in 1940, he discussed Chandrasekhar's work on white dwarfs. Figure 8.1 has been reproduced from this classic book.

Getting back to our story, Milne was not amused by Chandra appealing to the great physicists to come to his rescue. In a letter to Chandra, Milne wrote:

> Your marshalling of authorities such as Bohr, Pauli, Fowler, Wilson, etc., very impressive as it is, leaves me cold. If the consequences of quantum mechanics contradict very obvious, much more immediate, considerations, then something must be wrong with the principles underlying the derivation …To me it is clear that matter cannot behave as you predict …A theory must not be used as a doctrine, to compel belief …
>
> Eddington is nearly always wrong in his work in the long run, and I am quite prepared to believe that he is wrong here, in his details. But I hold by my general considerations.

As for Eddington, he carried on his tirade at various meetings. His last meeting with Chandra was in Paris in July 1939. Eddington tried to reach out to Chandra, but was not willing to change his stance. Soon World War II broke out, and Eddington died in 1944.

The astronomical community continued to ignore Chandrasekhar's seminal discoveries for more than three decades. By then Chandra was quite famous for his numerous *other* contributions. Many awards came his way. He was elected a *Fellow of the Royal Society* in 1944. The *Royal Astronomical Society* gave him the *Gold Medal* in 1952. A year later he received the coveted *Bruce Medal*. *But none of these awards mentioned his work on white dwarfs!*

Finally, things changed in the 1960s. This decade saw the discovery of *Binary X-ray Sources, Quasars and Neutron stars*. This Golden Age of great discoveries culminated with the discovery of the first stellar mass Black Hole in 1973. It became amply clear that **not all stars have sufficient energy to cool**.

In 1974, Chandra was given the *Dannie Heineman Prize*. And for the very first time, his work on white dwarfs was mentioned! Quite extraordinary, is it not? Curiously, by the time Chandra passed away in 1995, most astronomers remembered him almost exclusively for his monumental work on white dwarfs.

Chapter 9
Guest Stars

The Oriental Astronomers

Every now and then, the serenity and permanence of the sky is disturbed by the appearance in the sky of a new star. The most diligent observers of these new stars during the first two millennia were the oriental astronomers, notably in China, Japan and Korea. They not only observed them, but also recorded detailed description. Figure 9.1 shows the inscription on a Chinese *oracle bone* dating back to 1300 BC. Such oracle bones were made from an animal's shoulder blade, and often inscribed with a question. The inscription shown in the figure reads:

On the 7th day of the month a great new star appeared in the company of Antares.

Typically, these new stars could be seen for several months, sometimes even during the day! Occasionally they could be seen for a couple of years. The Chinese astronomers called them *guest stars;* like well-mannered guests, they left after a while! Chinese astronomers spent a great deal of effort looking for such guest stars. Such events were believed to foretell important earthly events such as the birth or death of a prince.

The Guest Star of AD 1006

Historical records tell us that the Chinese recorded six guest stars during the first and second millennium: AD 185, 386, 393, 1006, 1054 and 1181. By far the brightest of them was the Guest star of 1006. The most vivid description of this is by the Egyptian *Ali Ridwan*, who lived in Cairo. Ali recorded the position of this star, as well as the exact positions of the planets at the time of the first sighting of this star. He described it thus:

G. Srinivasan, *Life and Death of the Stars*, Undergraduate Lecture Notes in Physics,
DOI: 10.1007/978-3-642-45384-7_9, © Springer-Verlag Berlin Heidelberg 2014

Fig. 9.1 A Chinese *oracle bone* dating back to 1300 BC with a description of a *guest star.* This is the oldest known record of a *guest star.* Figure reproduced from, *Science and Civilization in China*, by J. Needham, Cambridge University Press

> It was a large nayzak, round in shape and its size two-and-a-half or three times the size of Venus. Its light illuminated the horizon and it twinkled a great deal. It was a little more than a quarter of the brightness of the Moon.

The Chinese, too, had seen this star. According to their recordings, 'it illuminated the horizon', 'it cast shadows', 'objects could be seen in its light' and so on. Clearly, it was a very bright star, and it could be seen for several years.

The Guest Star of AD 1054

Perhaps the most famous of the guest stars was seen on July 4 1054 (anticipating America's Independence Day by several centuries?!). This was in the constellation *Taurus*, the Bull; the Bull being chased by *Orion,* the Hunter. It is clear from the detailed description that a month before it reached its maximum brightness it was as bright as Jupiter, and was visible during daylight for 23 days after it reached maximum brightness. It eventually faded away 630 days later.

This particular Guest star will be the focus of our attention in the next book in this series, *Neutron Stars and Black Holes*.

De Nova Stella of AD 1572

The Guest star of AD 1572 was seen and described by many European astronomers and mathematicians. But the most comprehensive study of this new star, *Nova Stella* (in Latin), was by another rising star—this time in the world of astronomy. He was the young Danish astronomer by name **Tycho Brahe,** shown in Fig. 9.2. As **Johannes Kepler**, later to be his pupil, said, 'If that star did nothing else at least it announced and produced a great astronomer'.

Tycho Brahe gathered all his observations of this guest star in his famous book, *De Nova Stella,* published in 1573. Here is his description of the first seeing:

> Last evening in the month of November, on the eleventh day of that month, in the evening, after sunset, when according to my habit, I was contemplating the stars in a clear sky, when I noticed that a new and unusual star, surpassing the other stars in brilliancy, was shining almost directly above my head; and since I had, almost from boyhood, known all the stars of the heavens perfectly, it was quite evident to me that there had never before been any star in that place in the sky, even the smallest, to say nothing of a star so conspicuously bright as this.

Using his homemade instrument, he measured the position of the star very accurately. And he did this as often as he could. The purpose, of course, was to detect any motion. After eighteen months of painstaking observations he concluded that there is no motion. This ruled out the possibility that the new star might be associated with a planet. To quote Tycho Brahe:

> I conclude therefore that this star is not some kind of comet or a fiery meteor … but that it is a star shining in the firmament itself.

One of the very important things that Tycho Brahe did was to measure the brightness of the new star at regular intervals, by comparing its brightness at any given time with the brightness of other known stars in the sky. And he did this very meticulously. Since the brightness of the standard stars are unlikely to have changed in the intervening four centuries, one can use Tycho Brahe's description to derive what is known as the *light curve* of the new star of 1572. Such light curves—*a plot of the variation of the brightness with time*—are now recognized to be crucial. Here is a sample of Tycho Brahe's description of the changing brightness of the new star:

> When first seen the nova outshone all fixed stars, Vega and Sirius included. It was even a little brighter than Jupiter.
>
> The nova was as bright as Venus in November [1572]. In December, it was about equal to Jupiter. In January [1573] it was a little fainter than Jupiter and surpassed considerably the brighter stars of the first class. In February and March it was as bright as the last-named

Fig. 9.2 Tycho Brahe (AD 1546–1601)

> group of stars. In April and May it was equal to the stars of the second magnitude. After a further decrease in June, it reached the third magnitude in July and August … At the end of 1573 the nova hardly exceeded the stars of the fifth magnitude. Finally, in March 1574, it became so faint that it could not be seen any more.

Kepler's Nova Stella of AD 1604

Tycho Brahe died in 1601. But his research in astronomy was continued and carried forward by his pupil and assistant Johannes Kepler. In 1604, Kepler had the privilege of detecting another nova stella. And he studied it with the thoroughness that was the hallmark of everything he did. He, too, like Tycho had done, obtained a very accurate light curve. Unfortunately, all this was just before the advent of the astronomical telescope. As you know, the first astronomical observations with a telescope were done in 1609 by **Galileo Galilei**, observations that revolutionized astronomy.

The Guest Star in the Andromeda Nebula

Although we owe a lot to the oriental astronomers, and European astronomers like Tycho Brahe and Johannes Kepler, the true nature of the guest stars continued to be a big mystery. This was to change dramatically around 1930.

Fig. 9.3 The great spiral galaxy M31 in the constellation Andromeda (from the Wikimedia Commons, with the kind permission of the author, John Lanoue). Countless such spiral *nebulae* were thought to be part of our galaxy until Edwin Hubble established that this *nebula* was at a distance of nearly three million light years. Since our galaxy is only a hundred thousand light years across, M31 could not be in our galaxy. It had to be a galaxy in its own right!

The beginning of this chapter of the story goes back to 1885. One night in August 1885, the Russian astronomer E. Hartwig was entertaining some friends at his observatory. Some of them were curious to look through the telescope. So he decided to show them the Great Spiral Nebula M31 in the constellation Andromeda (see Fig. 9.3), an object which he observed regularly. He was astonished to find a bright new star near the centre of the nebula. Although he was absolutely sure that the star had not been there fifteen days earlier, he could not convince the Director of the observatory. He was allowed to announce this discovery only a week later after he and the Director had confirmed the existence of this new star. Hartwig followed the brightness of this new star for 180 days after its maximum brightness, after which it could not be observed. This nova was named *S. Andromedae*.

The Great Debate

At the dawn of the twentieth century, the nature of the large number of such spiral nebulae was not at all clear. Some felt that they were beyond our own Milky Way, while others argued that they were part of our galaxy. This debate gathered momentum with the discovery of more novae in spiral nebulae. By this time, astronomical

Fig. 9.4 The supernova that occurred in 1994 in the galaxy NGC 4526. This galaxy is at a distance of 55 million light years, compared to a *mere* 3 million light years in the case of the Andromeda galaxy. Notice that the brightness of the supernova (at the *bottom left*) is a significant fraction of the brightness of the central region of the galaxy. The *dark lanes* seen in this image are the spiral arms of the galaxy. Like in our own galaxy, there are a lot of *dust clouds* in the spiral arms. These opaque clouds are the gas clouds from which massive stars form, and which eventually explode as supernovae. [*Credit:* NASA, ESA, Hubble Key Project Team, and The High-Z Supernova Search Team]

photography had become more sophisticated. Several leading astronomers obtained photographic images of M31. These threw up several more novae in M31, but all of them were much fainter than S. Andromedae. It became clear that S. Andromedae was not a typical nova in M31. In 1917, the American astronomer **H. D. Curtis** noticed that the typical novae in M31 were roughly 10,000 times fainter than the typical novae in our own Galaxy. From this he concluded that M31 must be roughly 500,000 light years from us—a distance much larger than the size of our Milky Way Galaxy. This led Curtis to advance the so-called *island universes* hypothesis, which held that spiral nebulae were actually independent galaxies. But the distinguished astronomer **Harlow Shapley** at Harvard University strongly disagreed with this conclusion. In 1920, a formal debate took place between Shapley and Curtis under the auspices of the National Academy of Sciences in Washington, DC. But the great debate was inconclusive.

The stalemate was finally broken in 1923. Using the powerful 100-inch telescope at Mount Wilson in California, **Edwin Hubble** identified for the first time a few variable stars known as *cepheids* in M31. The point about these stars is that the distance to this class of stars could be uniquely determined by determining the period of waxing and waning of their intensity. This discovery enabled Hubble to determine the distance to M31—*The Great Andromeda Nebula was roughly 3 million light years from us.* This proved beyond all doubt that the Great Spiral Nebula in Andromeda was not part of our Galaxy. It was an entirely separate galaxy, containing a hundred billion stars.

A Super Nova?

The resolution of one problem threw up another one. The mystery of the typical novae in M31 being 10,000 times fainter than the novae in our Galaxy was finally solved—the Andromeda Galaxy was three million light years away. It was as simple as that. But S. Andromedae of 1885 now posed a very serious problem. *At its brightest, S. Andromedae was roughly one-sixth as bright as the entire galaxy.* Yes, one-sixth as bright as the light from a hundred billion stars! Since a photograph of S. Andromedae is not readily available, we have shown in Fig. 9.4 another example of a super bright nova in an external galaxy to illustrate how the light from such a nova can be a substantial fraction of the light from the entire galaxy.

By 1933, random photography of galaxies threw up many more examples of novae whose brightness nearly equalled the brightness of the host galaxy. This led the astrophysicist **Fritz Zwicky** at the California Institute of Technology to coin the phrase **supernova!**

Chapter 10
Supernovae, Neutron Stars and Black Holes

As was mentioned in the Chap. 9, it was Fritz Zwicky who christened the ultrabright *Novae Stella* as *Supernovae*. According to the folklore, he is supposed to have first used this phrase during one of his class room lectures in 1931. Zwicky estimated that in a supernova, such as S Andromedae of 1885, *the energy released during a period of a few weeks could equal what the Sun would radiate in a million years!* How is this energy produced? Zwicky and his colleague Walter Baade (one of the world's most distinguished observational astronomer) rejected the possibility that this staggering amount of energy is produced by the same process that makes the Sun shine. But remember that in 1931 one did not know the details of how stars generated energy; that great problem was solved only in 1938 by Hans Bethe. At the time we are talking about, there was just the conjecture by Eddington that stars generated their energy by converting hydrogen to helium.

The Discovery of the Neutron

The year is 1931. Let us go over to Cambridge to see what is happening there. Lord Rutherford was very concerned with the general difficulty in reconciling Bohr's theory of *electrons rotating around the proton-filled nucleus,* and the *isotopes of elements*. According to Bohr's model of the atoms, the number of positively charged particles inside the nucleus must be equal to the number of orbiting electrons. This means that once we specify the atomic charge, the atomic mass (which is essentially the mass of the nucleus) should be uniquely determined. But Rutherford and his colleagues had discovered that many elements had *isotopes.* All the isotopes of a given element had the *same atomic charge, but different atomic mass*. This meant that while the charge of the nucleus of all the isotopes of an element was determined by the number of orbiting electrons, the mass of the nucleus was not determined by the number of electrons; it varied. This is only possible if the nucleus contained *neutral particles* in addition to the protons. If the number of these neutral particles was different in the different isotopes, then that would explain why their nuclear masses

G. Srinivasan, *Life and Death of the Stars*, Undergraduate Lecture Notes in Physics, DOI: 10.1007/978-3-642-45384-7_10, © Springer-Verlag Berlin Heidelberg 2014

were different. Since no neutral particle was known in 1930, Rutherford was forced postulate the *neutral doublet—a bound pair of an electron and a proton*—although there was no experimental proof of this. Incidentally, this is exactly what Eddington did to make a helium nucleus out of four protons. He packed two electrons and four protons into the nucleus to make the net positive charge of the helium nucleus equal to two!

James Chadwick was to unlock the door to this basic problem in 1932. Chadwick was Rutherford's student at the University of Manchester, and moved with his master to the Cavendish Laboratory in Cambridge in 1919. He became Rutherford's trusted assistant during the period of intense creativity at the Cavendish Laboratory. In 1932, **Joliot** and **Curie** published the observation that alpha particles incident on Beryllium produced evidence of carbon and an intense 55 MeV 'gamma ray'. Chadwick and Rutherford immediately realized that this result must be wrong; the energy of the gamma ray was too high and they suspected that a *neutral particle* must be involved. Within weeks, Chadwick established the reaction

$$\boxed{{}^{9}\mathrm{Be} + {}^{4}\mathrm{He} \rightarrow {}^{12}\mathrm{C} + {}^{1}\mathrm{n}}$$

The particle on the right-hand side was a *neutral* particle, and Chadwick christened it the *neutron.* Chadwick determined the ratio of the mass of the neutron to that of the proton to be 1.0090 (the modern value is 1.0085). This neutral particle penetrated even lead. Chadwick received the Nobel Prize for this discovery in 1935. With this discovery, the list of *elementary particles* had grown to three: the *electron, proton* and the **neutron.** One was now in a position to explain the **isotopes** of the elements.

> The isotopes of an element had the same number of protons,
> but different number of neutrons in the nucleus.

The Origin of Supernovae

A year after the discovery of the neutron, Baade and Zwicky, shown in Fig. 10.1, published one of the most extraordinary papers in all of astronomical literature. They presented a joint paper at the December 1933 meeting of The American Physical Society held at Stanford University in California. The abstract of this talk was published in January 1934 in the Society's journal *The Physical Review*. This is now recognized to be one of the most prescient papers in the history of physics and astronomy. We have, therefore, reproduced below the abstract of this paper in its entirety:

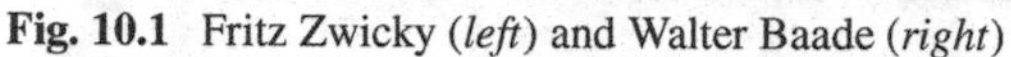

Fig. 10.1 Fritz Zwicky (*left*) and Walter Baade (*right*)

JANUARY 15, 1934 PHYSICAL REVIEW VOLUME 45
Supernovae and Cosmic Rays
by
W. Baade and F. Zwicky

Supernovae flare up in every stellar system (nebula) once in several centuries. The lifetime of a supernova is about twenty days and its absolute brightness at maximum may be as high as $M_{vis} = -14M$. The visible radiation L_v of a supernova is about 10^8 times the radiation of our Sun, that is, $L_v = 3.78 \times 10^{41}$ ergs/s. Calculations indicate that the total radiation, visible and invisible, is of the order of $L_T = 10^7 L_v = 3.78 \times 10^{48}$ ergs/s.
The supernova therefore emits during its life a total energy $E_T \geq 10^5 L_T = 3.78 \times 10^{53}$ ergs. If supernova initially are quite ordinary stars of mass $M < 10^{34}$ g, E_T/c^2 is of the same order as M itself. In the *supernova process mass in bulk is annihilated.* In addition the hypothesis suggests itself that *cosmic rays are produced by supernovae.* Assuming that in every nebula one supernova occurs every thousand years, the intensity of cosmic rays to be observed on the earth should be of the order of $\sigma = 3 \times 10^{-3}$ erg/cm^2 s. The observational values are about $\sigma = 2 \times 10^{-3}$ erg/cm^2 s (Millikan, Regener).

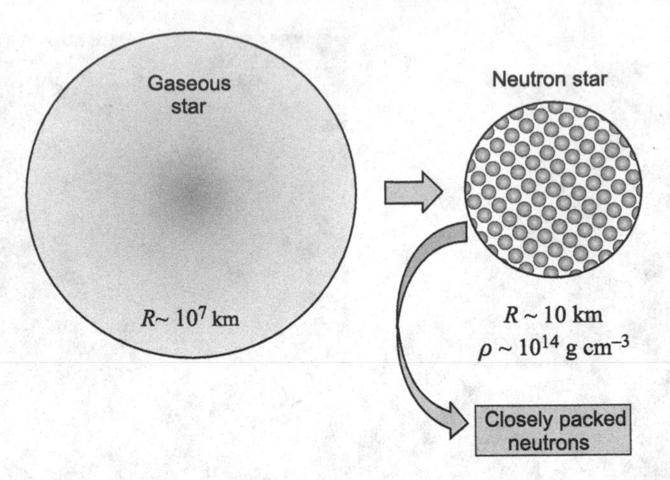

Fig. 10.2 A gaseous star of a few solar masses and radius of the order of 10 million km *implodes* and becomes a *neutron star*. Since a star is essentially made of hydrogen, initially the star consisted of roughly 10^{57} protons and equal number of electrons. According to Baade and Zwicky, somehow all these protons got converted to neutrons! A neutron star is just like a gigantic atomic nucleus, with the neutrons practically touching one another. Under such conditions, the density of matter would be 10^{14} g cm^{-3}. This is, indeed, the density of atomic nuclei that we are made of!

> With all reserve we advance the view that supernovae represent the transitions from ordinary stars into *neutron stars*, which in their final stages consist of extremely closely packed neutrons.

This paper is remarkable for the *density of brilliant ideas!*

- It asserts for the first time the existence of supernovae as a distinct class of astronomical objects.
- It introduces for the first time the name *supernovae*.
- It estimates correctly the total energy released in a supernova, although the reasoning is wrong. One may say that they got the answer right for the wrong reason!
- It gives a theoretical scenario for how cosmic rays are produced.
- It invents the concept of *neutron stars*.
- It suggests that *supernovae represent transitions of ordinary stars into neutron stars*.

I hope you are impressed! Let us try to understand the last point first. Their basic idea is explained in Figs. 10.2 and 10.3.

Baade and Zwicky realized that not only an enormous amount of energy had to be released; it had to be released in a short period of time. This ruled out the standard processes that produce energy in a steady manner in the stars; they did not know what these processes were, but it did not matter to them. One way to release energy quickly is *if the star were to suddenly collapse to a small radius*. Let us say that the

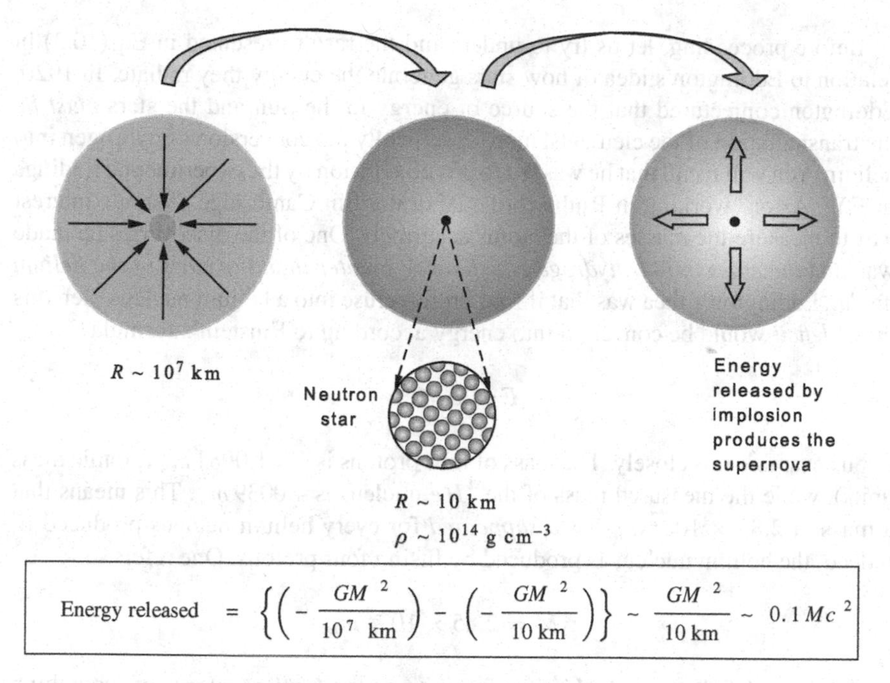

Fig. 10.3 When a neutron star of roughly a solar mass is formed, the gravitational potential energy released will be of the order of 10 % of its rest mass energy! Baade and Zwicky conjectured that this energy release causes the supernova explosion

original radius of the star was ~10 million km, and the final radius is ~10 km. The gravitational potential energy that would be released in the process would be

$$\Delta E = (\text{PE})_{\text{initial}} - (\text{PE})_{\text{final}} \,. \tag{10.1}$$

Recall that the gravitational potential energy of a mass M with radius R is

$$\text{Gravitational P.E.} \sim -\frac{GM^2}{R} . \tag{10.2}$$

Therefore, the energy released during the implosion of the star is

$$\text{Energy released} = \left\{ \left(-\frac{GM^2}{10^7\,\text{km}} \right) - \left(-\frac{GM^2}{10\,\text{km}} \right) \right\} \sim \frac{GM^2}{10\,\text{km}} \sim 0.1 Mc^2 . \tag{10.3}$$

The last step in Eq. (10.3) can be verified by multiplying and dividing GM^2/R by c^2, and assuming that $M \sim M_\odot$. Therefore, *the gravitational binding energy released is roughly 10 % of the rest mass energy of the neutron star*. It is the energy you would get if you **annihilated 10 % of the mass of the star**!

Before proceeding, let us try to understand the result presented in Eq. (10.3) in relation to Eddington's idea of how stars generate the energy they radiate. In 1920, Eddington conjectured that the source of energy in the Sun and the stars *must be* the transmutation of the elements, more specifically the conversion of hydrogen into helium. You will recall that he was led to this conclusion by the experimental findings of F.W. Aston, working in Rutherford's laboratory in Cambridge. Aston's interest was to measure the masses of the atoms accurately. One of the discoveries he made was that *the mass of four hydrogen nuclei was greater than the mass of the helium nuclei*. Eddington's idea was that if four protons fuse into a helium nucleus then this *mass deficit* would be converted into energy according to Einstein's formula

$$E = \Delta M c^2$$

Let us examine this closely. The mass of four protons is $4 \times 1.0081\, m_u$ (atomic mass units), while the measured mass of the ^4He nucleus is $4.0039\, m_u$. This means that a mass of $2.85 \times 10^{-2}\, m_u$ has *disappeared* for every helium nucleus produced if, indeed, the helium nucleus is produced by fusing four protons. One refers to

$$\Delta M = 2.85 \times 10^{-2}\, m_u$$

as the *mass deficit*. $E = \Delta M c^2$ is referred to as the *binding energy* of the helium nucleus. This *binding* is due to the *nuclear force*, the force that holds the nucleus together. If it were not for this strong binding, the nuclei of atoms would break up due to the coulomb repulsion between the protons. This mass deficit is roughly 0.7 % of the original mass of hydrogen (that is, the mass of four protons), and corresponds to an energy of about 26.5 MeV. This is the energy released when one helium nucleus is fused together. Conversely, if we want to break apart a helium nucleus, this is the energy we would have to *spend*. If mass M of hydrogen is converted into helium, then the energy released is $0.007\, Mc^2$. The mass of the sun is $2 \times 10^{33} g$, most of it hydrogen. By converting most of it to helium, it can generate $\sim 10^{52}$ erg of energy. The rate at which it radiates this energy (its luminosity) is 4×10^{33} erg/s. Therefore, the sun can easily shine for 10^{11} years by tapping this source of *subatomic energy*. This was Eddington's idea.

The energy released in the formation of a neutron star is again the binding energy. But this time the binding is due to *gravity*. One can speak of a mass deficit in this case also. The mass of the resultant neutron star, measured in terms of its gravity, is less than the sum of the masses of the neutrons. This mass deficit is roughly 10 % of the sum of the masses of the neutrons. The mass deficit when four protons are fused together is only 0.7 % of the mass of the four protons. Therefore, the energy released in the formation of a neutron star, per gram of matter, is much larger than the energy released in fusion reactions. Basically, this was the idea of Baade and Zwicky. And if this energy could be released in a short time, then one would have explained the energetics of a supernova.

You might be somewhat surprised by the above conclusion that the gravitational binding energy released during the formation of a neutron star far exceeds the nuclear binding energy. After all, gravitational force is supposed to be 10^{40} times weaker than the strong nuclear force! This is, of course, true. When we are dealing with a few particles inside a nucleus, the nuclear force dominates over the gravitational force. But when we are dealing with a gigantic nucleus consisting of 10^{57} particles, gravitational force is stronger by a huge margin. When we are at school, we are told that the Sun is the ultimate source of energy on Earth. In a similar fashion, in most astronomical situations *gravity is the ultimate source of energy.*

Now that we understand how Baade and Zwicky produced the energy observed in a supernova, let us now go to the next step in their argument. If a star like our Sun, with a radius of a million km and a mean density of 1.4 g cm^{-3}, collapsed to radius of 10 km, the density of the resultant star would be a few times 10^{14} g cm^{-3} (since density is inversely proportional to R^3, a decrease in the radius by a factor of 10^5 would mean an increase in density by a factor of 10^{15}). Such an incredible density may sound ridiculous to you. But we are all made of atomic nuclei whose density is $\sim 2.5 \times 10^{14}$ g cm^{-3}. Perhaps you did not know this. Take your favourite element from the Periodic Table, and estimate the density of the *nucleus* by dividing its mass by its volume. You will get the above number! Do convince yourself of this. The fact that our mean density is close to that of water—which is why we float in water—is simply because the mean distance between the ultra dense nuclei is very large. The mean distance between the *atoms* is $\sim 10^{-8}$ cm, while the size of the nuclei is $\sim 10^{-13}$ cm. The next step is easy. If you pack the star into a small sphere with nuclear density, you will get a gigantic nucleus. A gigantic nucleus 10 km in radius, instead of 10^{-13} cm!

However, since hydrogen was the most abundant element in our original star, such a giant nucleus should consist mostly of *protons*. But Baade and Zwicky talked of a *neutron star* and not a proton star. They not only did not give any reason for this, they are silent on this! *How and why did the protons transform themselves to neutrons?* You might say that this is a minor matter. Whether the result of the implosion is a neutron star or a proton star, its mass and size would be the same. And, therefore, the gravitational potential energy released would be the same. Hence, as for as the origin of supernovae is concerned, it really does not matter if it is a neutron star or a proton star.

Having discussed the brilliant ideas in the historic paper by Baade and Zwicky, let us also take stock of some of its weak points.

1. As mentioned above, they do not give any reasons for why the result of the implosion would be a *neutron star.* Indeed, the physics underlying the neutronization of matter had not yet been discovered in 1933.
2. Baade and Zwicky do not give any reason for why a star would implode. One possibility is, of course, that the star was a failed white dwarf. If its mass exceeded the *Chandrasekhar Limiting Mass,* it would have no option but to collapse beyond the white dwarf stage. *But there is no reference in their paper to Chandrasekhar's seminal discoveries!*

3. Nor do they give any argument for the timescale in which such a collapse would take place. This was crucial for explaining the supernova phenomenon. The details of such an implosion became clear only in the 1960s.

So the paper by Baade and Zwicky was a *shot in the dark* in many respects. But the important thing is that their extraordinary predictions have now been observationally verified. We now know that supernovae signal the birth of neutron stars! Baade and Zwicky might have been 'right for the wrong reasons' with regard to the details. **But they were right**!

Neutronization of Matter

As mentioned above, Chadwick discovered the neutron in 1932. At that time the nature of the forces that hold the nuclei together was ill understood. A formal theory of radioactivity, also known as beta decay, was discovered by Enrico Fermi only in 1934. Let us briefly recall the basic idea of Fermi's theory. In beta decay, radioactive nuclei emit β rays or electrons. Fermi explained that this is due to neutrons inside the nucleus decaying to protons, electrons and antineutrinos.

$$n \rightarrow p + e^- + \bar{\nu}$$

You may recall that Pauli had postulated that a *neutral particle* must be emitted in such a decay. After Chadwick had discovered the heavy neutral particle, which he called the *neutron*, Fermi christened the light neutral particle emitted in beta decay as *neutrino*. The neutrino was a central character in our story about energy generation in the Sun. The electron and the neutrino do not exist inside the nucleus. According to Fermi, they are spontaneously created when a neutron decays, just as photons are spontaneously created when an electron jumps from a higher energy level to a lower energy level in an atom. Fermi was to make many profound contributions to physics, both theoretical as well as experimental. His theory of beta decay was perhaps his most important theoretical discovery. Interestingly, his discovery paper was rejected by the prestigious British journal *Nature* because it considered the paper, *too remote from reality!* Fermi's paper was published by the German journal, *Zeitschrift für Physik* in 1934, and finally published by *Nature* five years later, after Fermi's work had been widely accepted.

In the radioactive decay we have been discussing, the charge of the nucleus increases by one since a neutron is converted to a proton (with the electron and the neutrino escaping):

$$(A, Z) \Rightarrow (A, Z + 1) + e^- + \bar{\nu}.$$

In the above reaction, A is the atomic mass and Z is the atomic charge. There is another type of beta decay, known as *inverse beta decay* or *electron capture*. The

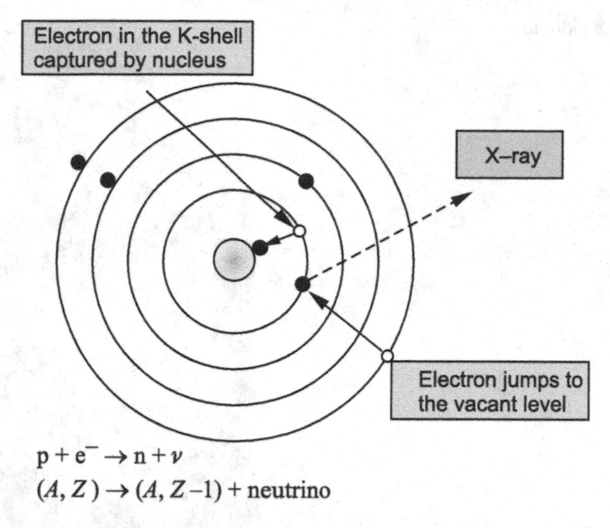

$$p + e^- \rightarrow n + \nu$$
$$(A, Z) \rightarrow (A, Z-1) + \text{neutrino}$$

Fig. 10.4 Inverse beta decay or *electron capture*. An electron in the K-shell of a neutral atom is swallowed by the nucleus. A proton inside the nucleus interacts with this electron and transforms itself into a neutron. The neutrino emitted during this reaction escapes. What is actually observed is a soft x-ray photon emitted when an outer electron jumps to the vacant state in the K-shell

theory of electron capture was first discussed by Giancarlo Wick soon after Fermi discovered the theory of beta decay. Wick was one of the many young brilliant students working with Fermi in Rome. The electron capture process is explained in Fig. 10.4. You will remember from your atomic physics course that the innermost electronic shell, known as the K-shell, can accommodate two electrons. *In heavy elements, the nucleus gobbles up one of the K-shell electrons, and a proton inside the nucleus is converted to a neutron.* The neutrino emitted in this process escapes. Noticing that there is a *vacancy* in the K-shell, an electron in one of the higher levels jumps to the K-shell. The energy difference is emitted as a soft x-ray. It is quite difficult to detect soft x-rays since they are easily absorbed. But they were eventually detected in 1937 by Luis Alvarez, confirming the ideas on electron capture. He first succeeded in detecting soft x-ray in vanadium-48, and from other heavy elements subsequently.

These ideas of Fermi and Wick were extended by the physicist Hund in 1936. He pointed out that this process of *inverse beta decay* would occur even if the electrons were not 'bound electrons'. In other words, if we have a Fermi gas of protons and electrons—like one has in a white dwarf—the inverse beta decay would occur *provided the density was sufficiently large.*

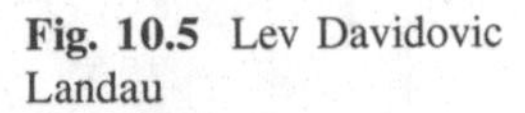

Fig. 10.5 Lev Davidovic Landau

Neutron Cores of Massive Stars

The above-mentioned ideas formed the basis of a most remarkable paper entitled, *Origin of Stellar Energy,* by the most distinguished Russian theoretical physicist **Lev Landau** published in 1938 in the journal *Nature* (Fig. 10.5). This prescient paper was forwarded to *Nature* by none other than Neils Bohr. But that is a different story.

The notion of a neutron star, with proper theoretical ideas to support it, can be traced to this paper which was *barely half a page long*! When Landau wrote this paper, the origin of stellar energy was still a mystery; Hans Bethe was to solve that problem only later that year. Landau invented neutron stars to solve this problem. There are two parts to Landau's paper. In the first part, he argues that at very high densities it would be energetically more favourable for matter to exist in a *neutronic state.* He then goes on to argue that if such neutron cores existed inside stars, it would be straightforward to explain the sustained luminosity of the Sun for several billion years. Let us first try to understand the idea of a neutronic state (see Fig. 10.6).

As we know, matter consists of nuclei and electrons. This is the kind of matter that we are familiar with. Landau called it *electronic state of matter.* In our discussion so far of gaseous stars and white dwarfs, we have assumed—and quite correctly—that stellar matter is also of the *electronic type.* As we have seen in the earlier chapters, the electrons become degenerate when the density becomes high. And because the degeneracy pressure can be immense, matter becomes quite incompressible. This is why white dwarfs are stable. Landau argued that if electrons combine with the nuclei to form neutrons, then the resultant matter would be much more compressible and, therefore, can attain higher densities. The end result of this process would be

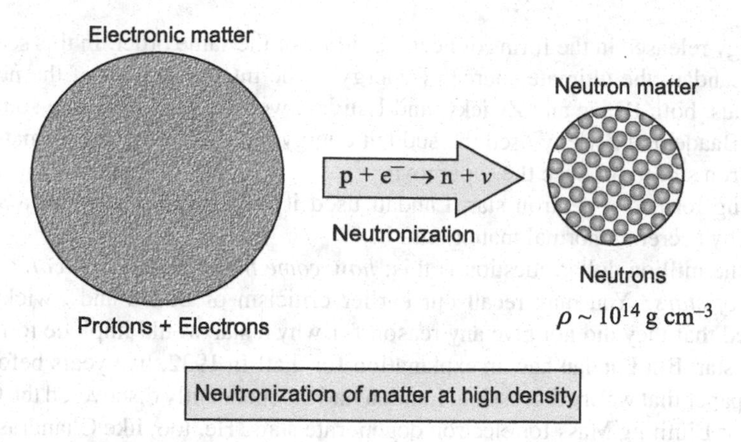

Fig. 10.6 This explains Landau's idea of how matter becomes neutron rich at densities beyond a critical density $\sim 10^{11}$ g cm^{-3}. Although this process is *endothermic*, the gain in gravitational potential energy due to the contraction of the sphere will compensate for the energy *spent* in the neutronization

a *degenerate neutron gas*, in which all the nuclei have combined to with electrons to form neutrons. It is easy to see why such a matter would initially be *soft* or *compressible*. As electrons disappear, the degeneracy pressure of the electrons will decrease (since this pressure is determined by the density). As a consequence, the compressibility of the matter would increase. It is true that since neutrons also obey Fermi–Dirac statistics they, too, will exert pressure, but as we have argued earlier, the pressure of nonrelativistic neutrons (or protons) will be 2000 times smaller than the pressure due to the electrons because of the greater mass of the neutrons. The pressure of the neutrons will become appreciable only when the density reaches $\sim 10^{14}$ g cm^{-3}. When this density is attained, neutronic matter will also become incompressible and stable.

But there is a catch in what we have just said. The reaction $p + e^- \longrightarrow n + \nu$ is a strongly *endothermic reaction*, that is, we have to supply energy for this reaction to take place. *To transform one gram of electronic matter into neutronic matter would cost us* 7×10^{18} erg. This is why the neutronic state of matter is unfavourable under normal conditions. It is good thing, too! Otherwise, atoms as we know them will cease to exist. And we, too, will cease to exist. But Landau was terribly clever. He argued that *when the mass of the body becomes very large, the gravitational energy gained in going over to the neutronic state compensates for the loss in internal energy.*

Let us now return to the second part of Landau's paper. Landau's main motivation was to invent a mechanism for sustained energy generation in the stars. Let us assume, for a moment, that every star has a neutron core. Atoms outside the core would fall in and will be accelerated to high speeds because of the enormous surface gravity of the neutron core ($GM_{\text{core}}/R_{\text{core}}^2$). When these atoms impact on the surface of the neutron core, their enormous kinetic energy would be converted to heat. Since the kinetic energy at the time of impact would be roughly 10 % of the rest mass energy Mc^2, the energy released in the forms of heat would be of the same order. In this scenario due to Landau, the ultimate source of energy is the intense gravity of the neutron core. Thus, both Baade and Zwicky, and Landau, were tapping the same source of energy. Baade and Zwicky used the sudden energy release during the formation of the neutron star to produce the supernova.

Having formed a neutron star, Landau used it to generate energy in a steady manner by accreting normal matter onto it.

But the million dollar question is this: *how come there are neutron cores at the centres of stars?* You may recall our earlier criticism of Baade and Zwicky. We remarked that they did not give any reason for why a star would implode to form a neutron star. But Landau had an explanation for that! In 1932, five years before his *Nature* paper that we are discussing, Landau had independently discovered the Chandrasekhar Limiting Mass for electron degenerate stars. He, too, like Chandrasekhar had done two years earlier in 1930, obtained the limiting mass to be about $1.5\,M_{\odot}$. Landau clearly stated in that paper, 'for $M > 1.5\,M_{\odot}$ there exists in the whole quantum theory no cause for preventing the system from collapsing to a point'. Five years later, developments in physics enabled Landau to state that such collapsing objects would find equilibrium as *neutron cores*.

Let me make a couple of observations before passing on to the Chap. 11 of this remarkable story.

- In 1932, after having independently discovered the Chandrasekhar Limit, Landau *rejected it*! To quote from that paper:

  > As in reality such masses [$M > 1.5\,M_{\odot}$] exist quietly as stars and do not show any such ridiculous tendencies we must conclude that all star heavier than $1.5\,M_{\odot}$ certainly possess regions in which the laws of quantum mechanics (and therefore of quantum statistics) are violated.

  Even the great Landau made the same mistake as Eddington and Milne!
- Landau had no explanation for why a star like the Sun, with a mass less than the critical mass, should possess a neutron core. He recognized this difficulty in his 1938 paper.
- We *now* know that star generate their energy by the transmutation of hydrogen into helium. Eddington conjectured this in 1920, and Hans Bethe worked out all the details in 1938. But the final proof came only in the year 2000 with the resolution of the solar neutrino puzzle (You will find a detailed discussion of this in the first book in this series, ***What Are the Stars?***). Landau's mechanism, ingenious as it is, is not the correct answer for the origin of stellar energy. But, as we shall see in the next volume in this series, ***Neutron Stars and Black Holes***, Landau's mechanism described in Fig. 10.7 is the correct explanation in a different context! It is now well established that there are countless number of neutron stars in binary systems with gaseous companions. The strong gravity of the neutron star pulls matter from the companion star. And when this matter accretes on to the surface of the neutron star, the neutron star becomes a very powerful x ray source. It is now widely accepted that these x rays are produced by the very same mechanism described in Fig. 10.7.

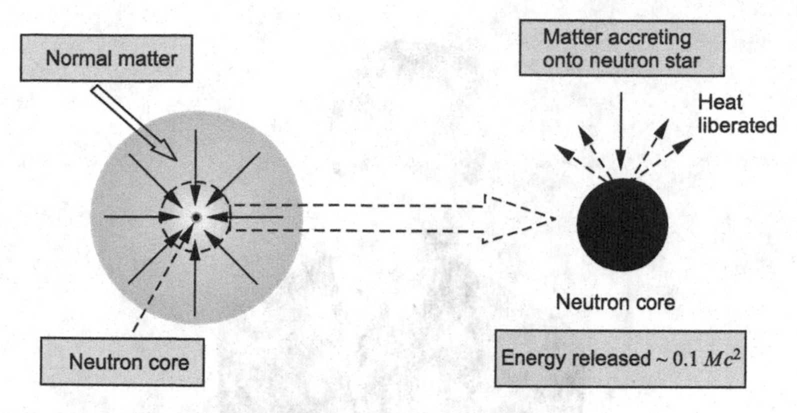

Fig. 10.7 Landau's mechanism for generating energy in stars

Interestingly, this idea was reinvented in 1964 by another Russian genius by the name Ya. B. Zeldovich.

The Maximum Mass of Neutron Stars

This remarkable paper by Landau attracted the immediate attention of another brilliant physicist. He was **Robert Oppenheimer** (shown in Fig. 10.8), a professor at the University of California at Berkeley, and also at CalTech. Oppenheimer was intrigued by Landau's estimate that a neutron star could have a mass as small as 0.001 $M_{\odot}$. He and his research associate Robert Serber thought about it very hard and came to the conclusion that Landau had got it wrong! After writing up their result and sending it to the *Physical Review*, Oppenheimer started thinking about a different question. Landau wanted to know what the *minimum mass* for a neutron star is. Oppenheimer wanted to know what the *maximum mass* was. I shall now explain how Oppenheimer and his student Volkoff answered this question.

The stability of a neutron star should be understood in a manner very similar to that of a white dwarf. As we saw, in a white dwarf the inward pull of gravity is balanced by the degeneracy pressure of the electrons. *In a neutron star, gravity is balanced by the degeneracy pressure of the neutrons*. Although the pressure due to the neutrons is negligible at white dwarf densities (because of their larger mass), by the time the density increases to 10^{14} g cm^{-3} the pressure due to the neutrons is sufficient to arrest gravity. Therefore, to construct models of neutron stars of various masses, we could simply apply Chandrasekhar's theory of white dwarfs to a neutron star.

If we were to adapt Chandrasekhar's theory, the stability of a neutron star is to be understood in terms of the equations,

Fig. 10.8 J. Robert Oppenheimer

$$\frac{dP}{dr} = -\frac{GM\rho}{r^2}, \tag{10.4}$$

where the pressure due to the neutrons is given by

$$P_{\text{deg}} = \text{K}_1 \rho^{\frac{5}{3}} \tag{10.5}$$

$$\text{K}_1 = \frac{1}{5}\left(\frac{3}{8\pi}\right)^{\frac{2}{3}} \frac{h^2}{m_n} \frac{1}{(\mu_n m_n)^{\frac{5}{3}}}. \tag{10.6}$$

Equation (10.4) is the equation of hydrostatic equilibrium we have encountered before in Eq. (6.2). The expressions (10.5) and (10.6) are identical in structure to Eqs. (6.10) and (6.11), respectively. The constant of proportionality K_1 in Eq. (6.11) for the pressure of an electron gas and Eq. (10.6) for a neutron gas differ slightly in two respects:

(i) The mass of the neutron replaces the mass of the electron in the denominator, and
(ii) The mean molecular weight per neutron μ_n replaces the mean molecular weight per electron μ_e. For electrons, we assumed μ_e to be 2. For the neutrons we have to set $\mu_n = 1$.

The latter point is easy to see. We introduced μ_e when we converted the number density of electrons into mass density. To do this, we had to multiply and divide by the *mass per electron*. Each atom contributes A heavy particles and Z electrons. Therefore, the mass of the material per electron is $(A/Z)m_p$. We argued that $(A/Z) \approx 2$ (except for hydrogen). Hence, the mass per electron is $2m_p$ and $\mu_e = 2$. In the case of a pure neutron gas, to convert the number density to mass density we simply have to multiply and divide by the mass of the neutron. In other words, $\mu_n = 1$.

Chandrasekhar's Theory for Neutron Stars

The mass–radius relation for neutron stars can be obtained in a manner identical to how we obtained this relation for white dwarfs (see the steps leading up to Eq. (6.15)). Not surprisingly, we shall get the same relation as Eq. (6.13) for white dwarfs:

$$\boxed{\begin{aligned} R &= \left(\frac{\mathrm{K}_1}{0.424G}\right)\frac{1}{M^{\frac{1}{3}}} \\ R &\propto M^{-\frac{1}{3}} \end{aligned}} \tag{10.7}$$

The constant of proportionality is, of course, different, as comparison of Eqs. (6.11) and (10.6) will reveal. Because of this the radius of a one solar mass neutron star will be much smaller than that of a white dwarf of the same mass. Using (10.6) and (6.11), the ratio of the radius of a neutron star to that of a white dwarf of the same mass can be written as

$$\frac{R(\text{neutron star})}{R(\text{white dwarf})} = \frac{(\mathrm{K}_1)_{ns}}{(\mathrm{K}_1)_{WD}} = \frac{m_e}{m_n}(\mu_e)^{\frac{5}{3}}$$

Remembering that the mass of the neutron is 2000 times the mass of the electron and $\mu_e = 2$, you can easily verify that the radius of a one solar mass neutron star will be roughly 15 km (instead of 10,000 km for white dwarf, and a million km for the Sun).

The 'Chandrasekhar Limiting Mass' for Neutron Stars

We saw that Chandrasekhar's original theory of white dwarfs was only approximate since he assumed the electrons to be nonrelativistic. Inclusion of the effects of Special Relativity led to a limiting mass, given by,

$$M_{Ch} = 0.197\left[\left(\frac{hc}{G}\right)^{\frac{3}{2}}\frac{1}{m_p^2}\right]\times\frac{1}{\mu_e^2} = 5.76M_\odot\times\frac{1}{\mu_e^2}.$$

For an assumed value of $\mu_e = 2$, we get the well known result $M_{Ch} = 1.4\,M_\odot$. In a similar manner, Chandrasekhar's theory if applied to a neutron star would predict a limiting mass for neutron stars given by the same expression, with $\mu = 1$:

$$\boxed{M_{Ch}(\text{neutron star}) = 0.197\left[\left(\frac{hc}{G}\right)^{\frac{3}{2}}\frac{1}{m_p^2}\right] = 5.76\,M_\odot} \tag{10.8}$$

At this limiting mass, the neutron star would be fully relativistic, and its radius would be zero.

Neutron Stars in General Relativity

If you are a careful reader, you would have noticed that Chandrasekhar's theory of white dwarfs was an exact theory in as far as the treatment of quantum statistics and the variation of mass with velocity predicted by the Special Theory of Relativity. But for the gravitational force, he assumed Newton's laws. This will be seen clearly in Eq. (10.4), for hydrostatic equilibrium.

The essential difference between Einstein's theory of gravity (which is described by the General Theory of Relativity) and Newton's theory is that in Einstein's theory all forms of energy contribute to gravity. Thus, the internal energy is also a source of gravity.

Having pointed this out, I should add that Newton's law of gravity is quite adequate when we are dealing with white dwarfs. It is true that in the relativistic electron gas the kinetic energy of the electrons is comparable to the rest mass energy of the electrons; that is what we mean by saying that the electrons are relativistic. But the internal energy of the electron gas is very small compared to the rest mass energy of the *nuclei,* which constitute the major part of the mass of the star. *In a white dwarf, therefore, gravitational effects are determined by the rest mass of the nuclei; the contribution to gravity from the energy of the electrons is very negligible.* This is why Newton's theory adequately describes gravity in a white dwarf.

But it is a different matter altogether when we are discussing a neutron star of large mass. There one expects the neutrons to be relativistic, just as the electrons were relativistic in a massive white dwarf. *In a relativistic neutron gas, the kinetic energy of the neutrons is comparable to the rest mass energy of the neutrons. Hence, the internal energy of the neutrons will contribute to gravity in a significant way.* If this is the case, the right-hand side of the equation of hydrostatic equilibrium (10.4) would have to be modified to take into account General Relativity.

Oppenheimer realized this, and made the necessary modifications. We shall not dwell on that here since we shall be discussing all this in detail in a subsequent volume devoted to *Neutron Stars and Black Holes.* To summarize, to derive the maximum mass of neutron stars Oppenheimer and Volkoff repeated Chandrasekhar's calculations, but with two modifications:

1. *They used the degeneracy pressure of the neutrons, instead of electrons.*
2. *They used Einstein's theory for the description of gravity.*

The rest was just tedious calculation, which young Volkoff carried out with great care and fortitude. Their conclusion was published in 1938 and is explained schematically in Fig. 10.9.

Oppenheimer and Volkoff drew the following conclusions from their work:

- The maximum mass of neutron stars is $0.7\, M_{\odot}$.
- The radius of the neutron star of this mass would be about 10 km, and
- The central density of a neutron star of maximum mass would be $\sim 5 \times 10^{15}$ g cm^{-3}.

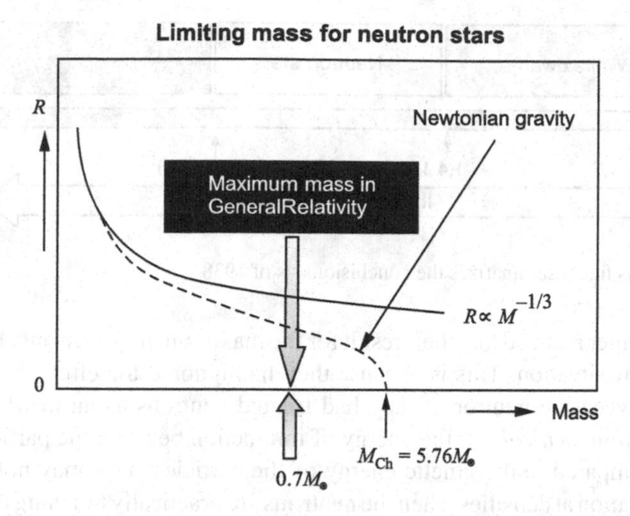

Fig. 10.9 Oppenheimer and Volkoff assumed that a neutron star consists essentially of neutrons. For the pressure of the neutrons they used the ideal Fermi gas equation of state as given by Chandrasekhar. The *solid curve* is the mass–radius relation they would have obtained on the assumption that the neutrons were nonrelativistic and Newtonian gravity. This relation would predict a radius of roughly 15 km for a neutron star of one solar mass. The *dashed curve* would be the neutron-star analogue of Chandrasekhar's exact theory of white dwarfs. This would predict a maximum mass of 5.76 $M_\odot$. A star of this mass would be fully relativistic and have zero radius. Oppenheimer and Volkoff modified Chandrasekhar's treatment of gravity to take into account the effects of General Relativity. According to their calculations, **the limiting mass of neutron stars is 0.7** $M_\odot$

There are several points about their result which deserve elaboration.

1. You may be surprised that a neutron star with the maximum mass has *finite radius*. In contrast, a white dwarf with a maximum mass of 1.4 $M_\odot$ has *zero radius*! At the *Chandrasekhar limit* for white dwarfs, the electrons would be fully relativistic; *all* the electrons would have speed almost equal to the speed of light. This is one of the reasons why the maximum mass of neutron stars is less than the *Chandrasekhar limit* for neutron stars, which is 5.76 $M_\odot$. A neutron star of mass equal to 5.76 $M_\odot$ would be fully relativistic and have zero radius, but in a neutron star of 0.7 $M_\odot$ the neutrons are only mildly relativistic. This is why it has a finite radius.
2. The reason why the maximum mass is less than 5.76 $M_\odot$ is that Oppenheimer and Volkoff treated gravity using General Relativity. As mentioned earlier, in General Relativity the internal energy also contributes to gravity. A smaller maximum mass is what we should expect, since gravity is *stronger* in General Relativity.
3. Since the neutron star has a finite radius at the *maximum mass,* one can ask the following question: *What happens if we increase the mass of the star beyond the maximum mass?* Put slightly differently, 'In what sense is it the maximum mass?' The answer is the following. For a star to be stable, *the central density should increase with increasing mass.* This condition is satisfied till we reach the *maximum mass,* but is violated beyond this mass. In other words, no stable stars are possible beyond the maximum mass. We shall defer a more careful discussion of this to the next volume in this series.

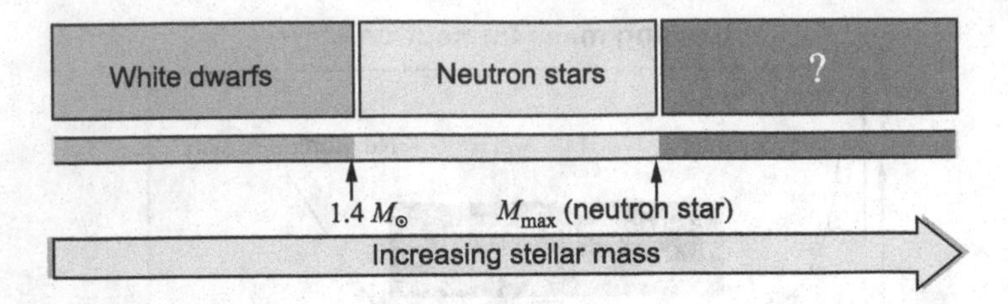

Fig. 10.10 This figure summarizes the conclusions as of 1938

4. Oppenheimer realized that their result for the maximum mass can only be regarded as an approximation. This is because they had ignored the effect of the nuclear force between the neutrons. They had treated neutrons as an *ideal Fermi gas*. Recall that in an *ideal gas* the energy of interaction between the particles is negligible compared to the kinetic energy of the particles. This may not be a good approximation at densities when the neutrons are practically touching one another. But in 1938, the nature of the nuclear force was not fully understood. It was not even clear whether it was attractive or repulsive at neutron star densities. Oppenheimer's intuition told him that at very short distances the nuclear force might be *repulsive* (we now know that he was right). If nuclear force is repulsive at very short distance, then it would help in supporting a star of a larger mass than what the degeneracy pressure alone was able to support. This intuition led Oppenheimer and Volkoff to *guess* that when proper account is taken of the nuclear force, the maximum mass of neutron stars might be *a few solar masses*. Seventy years later, we believe that the maximum mass of neutron star is around 2 solar masses.
   (By the way, there is an interesting point here that deserves comment. Chandrasekhar also assumed that the electrons in a white dwarf can be regarded an *ideal Fermi gas*. At the high densities that obtain in a white dwarf, one would expect the coulomb interaction between the electrons to be quite significant. Why did Chandrasekhar not worry about it? He did not have to! I shall tell you the answer now, and let you think about it. The reason why a high-density electron gas can be regarded as ideal is that *an electron gas has the peculiar property that it becomes more ideal as the density increases!* But this is not true of a neutron gas. We shall return to this in the next volume.)

The conclusions, as of 1938, concerning the ultimate fate of stars are summarized in Fig. 10.10.

Black Holes

The discovery of the maximum mass for neutron stars naturally led Oppenheimer to the question, *'What is the fate of massive stars that cannot find equilibrium as neutron stars?'* Like Chandrasekhar a few years earlier, Oppenheimer was also left speculating: *Either the Fermi equation of state must fail at very high densities, or that the star will continue to contract indefinitely never reaching equilibrium'.*

In 1939, Oppenheimer and Snyder (another brilliant student!) chose between these alternatives. They did this by carefully studying the implosion of a massive star. Of course, they had to make some simplifying assumptions about the star. They assumed that the star was spherical and nonrotating. Having made these assumptions, Snyder did the calculations in a *mathematically exact manner* within the framework of Einstein's General Theory of Relativity. And their conclusion was staggering! The best way to convey the impact of what they found is to quote the concluding sentences from their historic paper:

> When all thermonuclear sources of energy are exhausted a sufficiently heavy star will collapse. This contraction will continue indefinitely till the radius of the star approaches asymptotically its gravitational radius. Light from the surface of the star will be progressively reddened and can escape over a progressively narrower range of angles till eventually the star tends to close itself off from any communication with a distant observer. Only its gravitational field persists.
>
> Oppenheimer and Snyder, 1939

Put simply, the star will become a **black hole**!

This result of Oppenheimer and Snyder confirmed Chandrasekhar assertion of 1932. Let us recall that prophetic statement:

> For all stars of mass greater than $M_{critical}$ the perfect gas equation of state does not break down, however high the density may become, and the matter does not become degenerate. An appeal to the Fermi–Dirac statistics to avoid the central singularity cannot be made.
>
> S. Chandrasekhar, 1932

Oppenheimer and Snyder had confirmed Eddington's fear. Let us recall Eddington's speech at the eventful meeting of the Royal Astronomical Society in January 1935.

> The star has to go on radiating and radiating, and contracting and contracting until, I suppose, it gets down to a few km radius, when gravity becomes strong enough to hold in the radiation, and the star can at last find peace....
>
> Various accidents may intervene to save the star, but I want more protection than that. *I think there should be a law of Nature to prevent a star from behaving in this absurd way.*
>
> A. S. Eddington, 1935

As was mentioned earlier, Eddington and Chandrasekhar met for the last time in Paris in July 1939. Assimilating the work of Oppenheimer and Volkoff, Chandrasekhar concluded his talk at that conference as follows:

> If the degenerate cores attain sufficiently high densities (as is possible for these stars) the protons and electrons will combine to form neutrons. This would cause a sudden diminution of pressure resulting in the collapse of the star onto a neutron core giving rise to an enormous liberation of gravitational energy. This may be the origin of the supernova phenomenon.
>
> S. Chandrasekhar, 1939

This is where matters stood in 1939. Figure 10.11 is a grand summary of the spectacular conclusions we have discussed in the preceding chapters, conclusions arrived

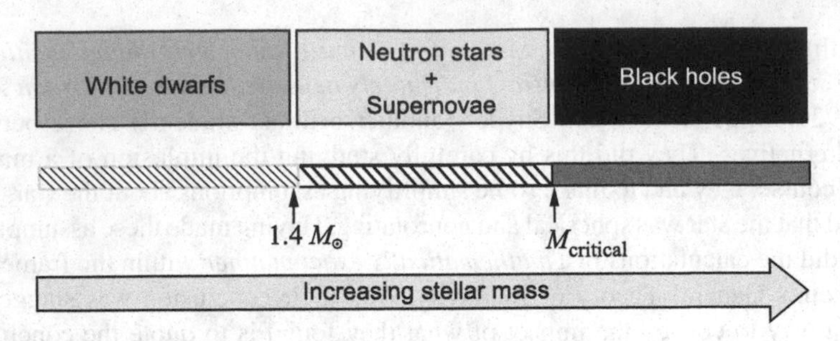

Fig. 10.11 The ultimate fate of stars. This was the scenario in 1939

at by *Fowler, Chandrasekhar, Baade and Zwicky and Oppenheimer* and his students *Volkoff* and *Snyder*.

Within weeks after the Paris conference World War II broke out. All the great physicists whose names we have encountered so far dedicated themselves to war efforts to defeat Hitler. The pursuit of science was interrupted for 6 years.

A Profile of Chandra

Subrahmanyan Chandrasekhar [1910–1995]

This volume of the series is being written during the Birth Centenary of Chandrasekhar. It is, therefore, appropriate that I include a brief sketch of his life and work. There is a tendency, particularly among non-scientists, to imagine scientists to be sterile and trivial personalities, totally devoid of any aesthetic sense. To dispel this, I have concentrated, in what follows, not so much on his science, but on his personality. Read on!

Subrahmanyan Chandrasekhar was a legend in his own time. When he passed away on 21 August 1995, prominent and detailed obituaries appeared in leading newspapers and magazines all over the world. Unfortunately, he was remembered mostly for his very early work on white dwarfs and the belated Nobel Prize he received, 53 years later.

But Chandra (as he was fondly known) was much more than a person who made a great discovery. He was one of the colossal figures of twentieth century science. Very few can match his sustained creativity and productivity for 65 years. His achievements have permanence in their character, and in his productivity and scholarship he has been compared with Lord Rayleigh and the great mathematician Henri Poincaré. As a mathematical physicist, he is regarded as one of the all time greats.

Chandra was born on 19 October 1910. He was born into a very cultured and gifted family. He burst into the international scientific scene at the young age of 18 years, when he was in the second year of his B.Sc. course in Presidency College, Madras. The year was 1928. In February 1928, C. V. Raman and his students, K. S. Krishnan among them, had discovered what has now come to be known as the *Raman Effect*. That summer, Chandra went to Calcutta to visit his uncle Sir C. V. Raman. The Indian Association for Cultivation of Science, where Raman had made the great discovery, was buzzing with excitement. A. H. Compton had just been awarded the Nobel Prize for the discovery of what is now known as the *Compton Effect*. There was expectation that Raman, too, would win the Prize. It was in this highly charged atmosphere that Chandra wrote his first scientific paper entitled the 'Thermodynamics of Compton scattering with reference to the interior of stars'. Soon after his return to Madras, the great German physicist Arnold Sommerfeld visited Madras. It was from him that Chandra learnt about the new developments in physics, in particular the discovery of the new statistics by Fermi and Dirac. Sommerfeld gave Chandra a copy of his paper in which he had used the new statistics to explain the behaviour of electrons in metals. Inspired by this, Chandra looked for another problem to 'apply the new statistics'. The newly discovered Compton Effect suggested an interesting problem. Within 2 months he had written a paper entitled, 'Compton Scattering and the New Statistics'. What is extraordinary is that he was so confident of the significance and correctness of his results that he sent the paper to R. H. Fowler in Cambridge requesting him to communicate it to the*Proceedings of the Royal Society*. Fowler did that, and the paper was published a few months later. Chandra was barely 18 years old at that time. And he never looked back after that. By the time he completed his degree in 1930, he had done his famous work on the theory of white dwarfs.

In 1930, Chandra went to Cambridge to work under the supervision of Fowler. We have already narrated this part of the story in Chaps. 6–8. The period between 1930 and 1935 was the most brilliant phase of his career. The papers he wrote during this period are now widely recognised to be at the base of the present revolution in astronomy. But unfortunately they were not seen so at that time. As we saw, the main reason was that Eddington did not believe the fundamental discovery Chandra had made. Faced with the enormous pressure of finding himself at the centre of a controversy with the leading astrophysicist in the world, Chandra decided to leave the subject of stellar structure altogether and move on to other things. He also decided to leave Cambridge. Just around that time he received an offer of a Research Associateship from Yerkes Observatory of the University of Chicago. The Director of the Observatory at that time was the distinguished astronomer Otto Struve. He was in the process of hiring some of the world's most brilliant astronomers and astrophysicists, and Chandra was one of them. Chandra stayed at the University of Chicago till he passed away.

Having decided to leave the subject of stellar structure, Chandra gathered together all his results and published a book entitled, *An Introduction to the Study of Stellar Structure*. This book is universally acclaimed to be masterpiece of the first rank. He was only 28 years old at that time.

Next he turned to the problem of the dynamics of star clusters. The novel way he approached the problems led to the birth of a new subject called Stellar Dynamics. In 1942, barely 4 years after the publication of his first book, he published his second book, *Principles of Stellar Dynamics*. One can say in retrospect that this was an unusual book in the sense that Chandra did not leave the field immediately. He continued to write a series of papers on the subject, some of them with the very famous mathematician John von Neumann, on the subject of the statistics of the gravitational field. It is in these papers that the seminal idea of *dynamical friction* was introduced and its consequences explored.

The next period 1943–1948 was devoted to an investigation of the extremely difficult problem of *radiative transfer* in stellar and planetary atmospheres. Incredibly, in that short span of time he managed to get exact analytical solutions to a large number of problems which had remained unsolved for nearly a century. Chandra often said that this phase of his career gave him the greatest satisfaction. His monumental book, entitled, *Radiative Transfer*, was published in 1950.

During the next decade he devoted his attention almost entirely to the difficult problem of the statistical description of turbulence, and hydrodynamic and hydromagnetic stability. He realised that unless substantial progress was made in these branches of physics, many interesting problems in astrophysics could not be approached. His mammoth book, *Hydrodynamic and Hydromagnetic Stability*, was published in 1961.

In the beginning of 1960s he was asked to give a series of four lectures at Yale University. He chose for the topic of his lectures, *Rotation of astronomical bodies*. While preparing for these lectures he became aware of the classic works of giants like Maclauren, Riemann, Jacobi and others. He realised, 'the subject, nevertheless, had been left in an incomplete state with many gaps and omissions, and some plain errors and misconceptions'. He devoted the next 6 years to clean up this extremely difficult field left incomplete by some of the greatest figures in the history of mathematics. The result was the publication of his book, *Ellipsoidal Figures of Equilibrium*, in 1969.

Around 1965 he got interested in General Relativity. The great revolution in *Relativistic Astrophysics* was just below the horizon. Chandra was very apprehensive about entering this field dominated by young brilliant stars like Roger Penrose and Stephen Hawking. The first problem he chose to attack suited his taste, talent and temperament. He went back to a problem he had worked on in Cambridge; he had started this work in collaboration with the great mathematician von Neumann. But the papers were never written. When World War II broke out, von Neumann got busy with war efforts in America. Chandra returned to this problem in 1964. There were two aspects to this problem: (1) What is the influence of General Relativity concerning the stability of stars? (2) Do dissipative effects of gravitational radiation induce instabilities in rotating stars? Both these questions were well posed, and were amenable to the kind of analysis in which he was the supreme master. This led to a series of papers which acquired great significance with the discovery of Pulsars, Quasars and Active Galactic Nuclei.

Next Chandra turned to the theory of *gravitational radiation*. As you are aware, in Newton's theory of gravity there is no gravitational radiation. Einstein and his collaborators had argued that gravitational radiation is a natural consequence of the General Theory of Relativity, just as the existence of electromagnetic radiation is a natural consequence of Maxwell's electrodynamics. But they could demonstrate this only with an approximate version of the full theory. So there were reasons to be cautious. Experience has taught us that very often the results obtained from an approximate version of a theory could be spurious. Therefore, not everyone was convinced about the existence of gravitational radiation. In 1964 Sir Hermann Bondi (one of the authors of the *Steady State Theory* of the Universe) wrote a classic paper in which he gave compelling arguments that gravitational radiation is a natural prediction of Einstein's General Theory of Relativity. The problem that Chandra set out to solve was the following. Since the General Theory of Relativity encompasses Newtonian theory as a limiting case, when the velocity of the bodies is small compared to the speed of light, one can try to develop Einstein's theory as a series expansion in powers of (v/c), with Newton's theory as the first term:

$$\begin{aligned}\text{General Relativity} \quad &= \text{Newton's Theory} + \text{terms of order } (\upsilon/c) \\ &\quad + \text{terms of order } (\upsilon/c)^2 + \cdots\end{aligned}$$

When $v \ll c$, one can drop all except the first term on the right hand side, and we recover Newton's theory of gravity. As the velocity increases, relativistic effects become more and more important. Consequently, one will have to retain more and more terms to have a satisfactory theory of gravity. When $v \sim c$, one will have to keep *all* the terms on the right-hand side. Such an approach is known as *post-Newtonian approximations*. Let us get back to gravitational radiation. Bondi had demonstrated that gravitational radiation exists in the exact theory of gravity. This raises an interesting question. Does gravitational radiation exist only in the full theory, or does it appear already in one of the post-Newtonian approximations to the full theory? Put differently, as we include terms of higher and higher order in (v/c), does gravitational radiation dramatically appear at some stage? It was this fundamental question that Chandra set out to answer with one of his students Yavuz Nutku. Within 2 years they were able to demonstrate the existence of gravitational radiation when one includes terms up to order $(v/c)^{5/2}$. This result had a tremendous impact on the general relativity community. Chandra was 60 years old then!

Just around that time relativistic astrophysics was coming of age. Penrose and Hawking had published their famous paper on the*singularity theorem*. Neutron stars had been discovered. Convincing arguments had been made that Quasars must harbour supermassive black holes. All this led to a revival of activity in general relativity. Some of the most brilliant students went to work on the physics of black holes. You will recall that Chandra's research career began with the study of white dwarves. Fundamental discoveries made by him in the early 1930s led one to the concept of black holes and singularities. It was therefore natural that Chandra, too, should enter this field. The problem he chose to concentrate on concerned the stability of black holes against external perturbations, such as electromagnetic waves and gravitational waves. Characteristically, after working on these problems for a number of years, and publishing a series of technically very difficult papers, he wrote a monumental book: *The Mathematical Theory of Black Holes*. This book was reviewed by Roger Penrose, the man who started the second revolution in relativity. Penrose ended his review thus: 'There is no doubt in my mind that this book is a masterpiece. It is clearly intended to last a long time. It will'.

Chandra was 75 years old at this stage. Many in the physics community were wondering what he would turn to next. Or would he take it easy and retire? He did neither of these, but continued with unabated enthusiasm. The problem he turned to next was the extremely difficult problem of collision of gravitational waves. When he decided to make this field his own, there were only two or three papers written on the subject; one of them by his former student Nutku and the others by Penrose. Very special assumptions had been made in these papers. Characteristically, Chandra wanted to solve this problem in all its generality. By 1988 he had done so!

Finally, at the ripe age of 80 years Chandra turned to the most difficult and the most ambitious project he had ever undertaken—to write a commentary on Newton's *Principia*. Like many, Chandra regarded this book as the greatest intellectual achievement of the human mind. Like everything else he started, he completed this project successfully. His book on Newton's *Principia* was published just a few weeks before he died.

Chandra began his research career at the age of 18 years. He sustained a very high level of productivity until he was 85 years old. During those 65 years or so, he wrote nearly 400 papers, none of them trivial and most of them significant. It is difficult to point to very many scientists who were creative for 65 years *at the limit of their abilities*. Sir Neville Mott and Hans Bethe are two names that come to my mind.

Chandra was a unique physicist even amongst the very great ones. *He was unique in his attitude to science, his quest for perspectives and beauty in science*. The most distinctive character of Chandra's scientific work was his attitude to science in general. As I have mentioned, there were seven periods in his life. He wrote six monumental books in which each subject was presented from a unified perspective, which was his own. About this attitude of striving to understand things in their own way, within his own framework, Chandra has written:

> 'After the early preparatory years my work has followed a certain pattern motivated, principally, by quest after perspectives. In practice, this quest had consisted of my choosing (after trials and tribulations) a certain area which appears amenable to cultivation and compatible with my taste, abilities, and temperament. And when after some years of study I feel that I have accumulated sufficient body of knowledge and achieved a view of my own, I have the urge to present my point of view '*ab initio*' in a coherent account with order, form and structure'.

Attaining complete understanding of an area, grasping and internalising it was the essence of Chandra's scientific life. To quote Chandra:

> 'If one's motivations are not galvanised to pursue science for its own sake, one's scientific life has not matured properly'.

Along with research, teaching was an integral part of Chandra's life. He prepared his classroom lectures with painstaking thoroughness, and they were delivered in a masterful way; every step of every argument was written on the board in his beautiful handwriting. More than 50 students worked with Chandra for their Ph.D. degrees. He considered his collaboration with young scientists an essential part of his scientific style. Indeed, he regarded his collaboration with young people more valuable than his collaboration with giants like von Neumann, Fermi and others! Young students who had the privilege of working with him benefitted enormously. To quote one of them: 'Chandra would transmit an enthusiasm, not in the ordinary sense that we will go and solve this or that difficult problem, but regarding how, in the end, after painstaking and lengthy calculations things would fall into place. Miraculous cancellations would occur and simple results would emerge'.

It is equally true that Chandra found it very inspiring to work with young people. This was particularly true after he got into General Relativity. He once said:

> 'I consider myself very fortunate in having made up my mind to do relativity. Among other things, for the first time, certainly after the early forties, I felt I was working in an area in which many others were far more equipped than I was. I thought I had a chance of having a close scientific proximity with people of the highest calibre. Certainly, to have known well and consider among my friends people like Roger Penrose, Stephen Hawking, Brandon Carter, Kip Thorne, James Bardeen—it is a marvellous experience, it is a kind of intellectual stimulation which I had not had before. Of course, I worked with Fermi. Fermi was a very great physicist, but here I am now in the community of young brilliant men. Even though in age I am very much elder than these people it has always been a satisfaction to me that these people treat me as their equal'.

This degree of genuine modesty is very rare indeed!

Chandra's writings have become legendary not only for their thoroughness, lucidity and scholarship, they also have a distinct style. Elegance and love for and attention to the language played as important role in his writings, as scientific facts and weaving them into mathematical formulae. Weisskopf, a very well-known physicist and who knew Chandra since his visit to Niels Bohr's institute in Copenhagen, Denmark in 1932, has said: 'He has an incomparable style. Good English style is a lost art in physics but he has it, and this wonderful feeling for the essential and a feeling for beauty'. In a similar vein, Lyman Spitzer of Princeton

University had remarked: 'It is a rewarding aesthetic experience to listen to Chandra's lectures and study the development of theoretical structures at his hands. The pleasure I get is the same as I get when I go to an art gallery and admire paintings'.

Chandra's deep interest in literature and classical music comes through in a transparent manner in his lectures and writings. To quote Weisskopf again, 'Right from the beginning, but even more later on, he became sort of the most pure example of the ideal scholar in physics... nothing of vanity, nothing of pushiness, nothing of job seeking, publicity seeking, or even recognition seeking...His deep education, his humanistic approach to these problems, his knowledge of world literature, and in particular English literature, are outstanding. I mean you would hardly find another physicist or astronomer who is so deeply civilised'.

An important aspect of Chandra's science was his quest for beauty in science. One may ask the question as to the extent to which the quest for beauty is an aim in the pursuit of science. He very seldom stated his own answers to such questions, but one may infer his views through his illustrations and examples of what other great scientists have responded to as *beautiful*. For example in a memorable lecture devoted to this question he quotes G. N. Watson's reactions to one of Srinivasa Ramanujan's incredible identities:

> '...such a formula gives me a thrill which is indistinguishable from the thrill I feel when I enter the Sagrestia Nuova of Capelle Medicee and see before me the austere beauty of *Day, Night, Evening and Dawn* which Michelangelo has set over the tombs of Giuliano de' Medici and Lorenzo de' Medici'.
> G. N. Watson

Chandra was fond of narrating what Werner Heisenberg thought was one of the truly momentous discoveries in the history of mankind:

> 'This was the discovery by Pythagorus that vibrating strings, under equal tension, sound together harmoniously if their lengths are in simple numerical ratios; in this discovery, for the first time, *profound connection between the intelligible and the beautiful was made*'.
> Werner Heisenberg

Those who have had the privilege of listening to Chandra's lectures, and reading his papers, will know that his concept of beauty in science was based on the following two criteria:

> There is no excellent beauty that hath not some strangeness in the proportions.
> Francis Bacon
> Beauty is the proper conformity of the parts to one another and to the whole.
> Heisenberg

That was Chandra! But all said and done, a scientist should be evaluated on the basis of his or her achievements. When Lord Rayleigh died, J. J. Thompson (who discovered the electron) gave the memorial address in the famous Westminster Abbey in London. He said:

> 'There are some great men of science whose charm consists in having said the first word on a subject, in having introduced some new idea which has proved fruitful; there are

others whose charm consists perhaps in having said the last word on the subject, and who have reduced the subject to logical consistency and clearness. Lord Rayleigh belonged to the second group'.

Chandra belonged to both groups! He and Rayleigh are perhaps the two greatest pillars of mathematical physics. But Chandra also had the privilege of saying the first word on a subject several times. He discovered:

1. The maximum mass of white dwarves.
2. That sufficiently massive stars cannot develop degeneracy, and will collapse to a singularity.
3. Dynamical friction in stellar systems.
4. Relativistic instabilities leading to gravitational collapse.
5. Gravitational radiation reaction.
6. Gravitational radiation-reaction-driven instability in rotating stars.

Chandra was himself very modest is assessing his own contributions.
Let me end this sketch of one of the truly great scientist by quoting Chandra himself:

> 'The pursuit of science has often been compared to the scaling of mountains, high and not so high. But who amongst us can hope, even in imagination, to scale the Everest and reach its summit when the sky is blue and the air is still, and in the stillness of the air survey the entire Himalayan range in the dazzling white of the snow stretching to infinity? None of us can hope for a comparable vision of nature and of the universe around us. But there is nothing mean or lowly in standing in the valley below and awaiting the sun to rise over Kanchenjunga.'
>
> S. Chandrasekhar

There is no doubt in my mind that posterity will regard Chandra as the most distinguished astrophysicist of the twentieth century.

Part II
The Life History of Stars: A Modern Perspective

Chapter 11
To Burn or Not to Burn

Nuclear Cycles

In Chap. 2 of this volume we discussed the *main sequence* of stars. In this phase of their lives, the stars are fusing hydrogen to helium in their cores. In stars less massive than the Sun, this proceeds via the *proton–proton chain*, while the *CNO cycle* is the dominant mechanism in stars more massive than the Sun. What happens after the hydrogen in the core is exhausted? Details apart, the answer to this question is rather simple. When all the hydrogen in the core is exhausted, the star will be left with a helium core and one would expect the helium to fuse to form carbon. When all the helium in the core is exhausted, carbon should fuse to form oxygen and so on. This is shown schematically in Fig. 11.1.

The basic idea of the nuclear cycle is that the by product of one fusion reaction is the fuel for the next. In more colloquial terms, the *ashes* of one stage of *burning* will be the *fuel* for the next stage.

I mentioned earlier that the stellar drama has many *acts*. Figure 11.1 would suggest that it should be just a *one-act play*, with many *scenes*. Why is this not the case? To appreciate this, let us recall some of our earlier discussion concerning fusion reactions in the Sun (see Chap. 5, 'Energy Generation in the Stars', in ***What Are the Stars?***).

Quantum Tunnelling

The main obstacle to fusing two nuclei together is the strong coulomb repulsion at short distances. For the case of two protons colliding against each other, the height of the coulomb barrier is ~1MeV (see Fig. 11.2). To put it differently, when the distance between the two protons is comparable to their size, the Coulomb repulsion energy

G. Srinivasan, *Life and Death of the Stars*, Undergraduate Lecture Notes in Physics,
DOI: 10.1007/978-3-642-45384-7_11,

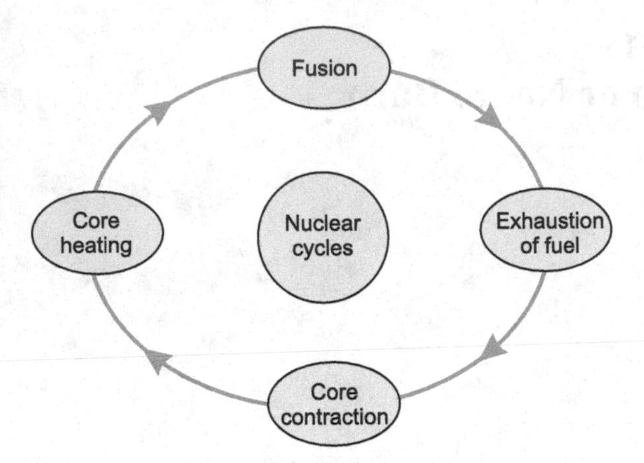

Fig. 11.1 Fusion reactions in stars proceed in cycles. The contraction of an inert core leads to heating. When the critical temperature is reached, the inert fuel will begin to fuse. When the fuel is exhausted, the core will, once again, become inert and consequently contract

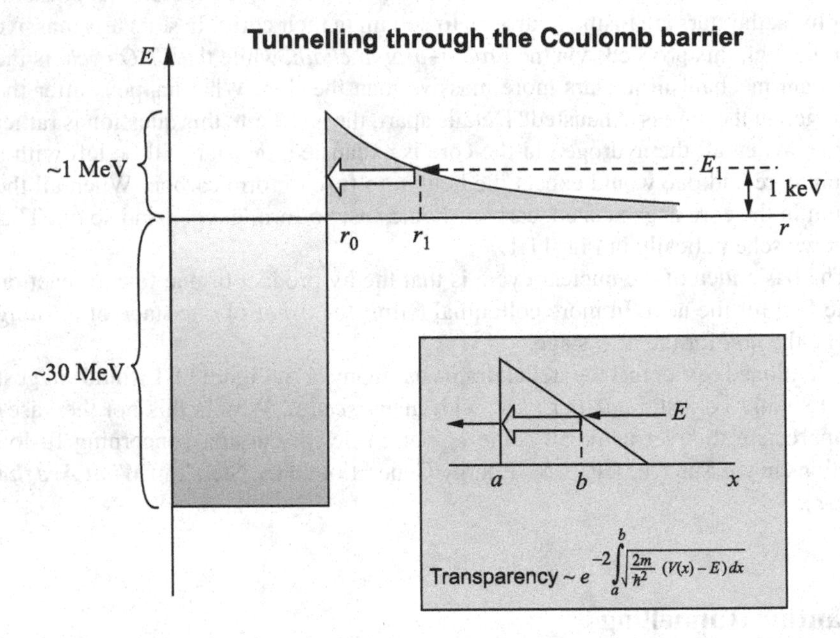

Fig. 11.2 For two protons to fuse together, they have to overcome the coulomb repulsion barrier. The height of this barrier is roughly 1 MeV. But the average energy of protons (at the central temperature of the order of 10^7 K) is only about 1,000 eV. Therefore, fusion is only possible due to quantum tunnelling through the coulomb barrier. The tunnelling probability is an exponential function

$$E_{\text{Coul}} = \frac{e^2}{r_0} \sim 1 \text{ MeV}, \tag{11.1}$$

where $r_0 \sim 10^{-13}$ cm. Do the protons near the centre of the Sun have enough energy to overcome this repulsion? Recall that the central temperature in the Sun is about 15 million degrees. The average energy of the protons is, therefore, ~1000 eV (remember that 10^4 K is of the order of 1 eV in energy units: $k_B \times 10^4$ K $\approx$ 1eV). This means that the typical energy of the protons is a *thousand times less than the height of the potential barrier,* which is ~1 MeV. A proton with energy ~1000 eV at infinity can never hope to climb the potential hill and fall into the hole at the centre. According to classical physics, it can only roll up the hill to a point where all its kinetic energy has been converted into potential energy (which is the point r_1 in Fig. 11.2); it is forbidden for the particle to temporarily *borrow* energy, climb up the hill and fall into the hole.

This great puzzle was solved in 1928 independently by the brilliant Russian physicist George Gamow, and by Condon and Gurney in the United States. The resolution of the problem invoked the newly emerging *quantum physics.* The underlying principle of quantum physics is the duality between particles and waves. It is this wave nature of particles that allows an alpha particle to escape from the nucleus. An analogy from optics (originally given by Gamow) will give us a feeling for how one may view this.

Imagine a beam of light incident on the boundary between two media at an angle greater than the *critical angle*. According to the laws of *geometrical optics,* we will have a total reflection of the incident beam—all the light will be reflected at the interface between the two media and no disturbance occurs in the second medium. However, if the same problem is treated within the *wave theory of light,* it is found that there is, in fact, some disturbance in the second medium as well. This is the phenomenon of *evanescent waves*; a phenomenon which is appreciable for a distance of the order of a few wavelengths of light. The evanescent wave decays exponentially as we go into the second medium. There is no interpretation of this disturbance which occurs in the second medium (which is predicted and measured by experiment) in the geometrical theory of light.

In the same manner, when we go from classical physics to quantum physics there is a possibility of particles penetrating potential barriers, or *tunnelling* through potential barriers. This possibility arises due to the wave nature of particles in quantum physics. Soon after the discovery of the theory of alpha decay by Gamow, and independently by Condon and Gurney, the transmutation of elements by proton capture was considered by Atkinson and Houtermans in 1929.

Given a potential barrier of a certain shape, the transparency or tunnelling probability can be calculated using wave mechanics. You will find the derivation in any introductory text on quantum mechanics for a variety of barrier shapes, such as triangular barrier, a rectangular barrier, the Coulomb barrier, etc. The transparency or the tunnelling probability through a barrier is defined as follows:

$$\text{Transparency} = \frac{\text{Transmitted Intensity}}{\text{Incident Intensity}}. \tag{11.2}$$

For an arbitrary barrier such as the one shown in the *inset* of Fig. 11.2, the transparency is given by

$$\boxed{\text{Transparency} \sim e^{-2\int_a^b \sqrt{\frac{2m}{\hbar^2}(V(x)-E)dx}}.} \tag{11.3}$$

It is worth noting the important features of the above expression for the tunnelling probability.

1. The tunnelling probability is an *exponential function.*
2. Given a barrier of a certain height V, *the tunnelling probability increases exponentially with increasing energy* of the incident particle.
3. The tunnelling probability *decreases exponentially with increasing thickness of the barrier.*
4. The probability of tunnelling is greater for particles of smaller mass.

So our success rate in fusing two protons together (the *reaction rate,* in the technical jargon) will depend upon an interplay between two opposite trends: *an exponentially increasing tunnelling probability with increasing energy* and *an exponentially decreasing fraction of particles with increasing energy* (recall the Boltzmann distribution).

A proper calculation for the case of the Coulomb barrier shows that the number of fusion reaction per unit volume and per unit time will involve an integral of the type:

$$J = \int_0^\infty e^{-E/kT} e^{-\eta/E^{\frac{1}{2}}} dE \tag{11.4}$$

The first exponential factor is the *Maxwellian tail* of the energy distribution and the second factor is the exponentially increasing tunnelling probability. The product of these two exponentials will give a *peak*, known as the *Gamow peak*. The area under this peak will determine the reaction rate. This is shown in Fig. 11.3.

Now let us get back to the nuclear cycles (Fig. 11.1). You will notice that between the exhaustion of fuel and the commencement of the next stage of fusion there are two important steps: *core contraction* and *core heating*. This is why the stellar drama has many *acts* and is not a one-act-drama. To understand this better, let us go back to the Coulomb barrier.

When two protons collide against each other the height of the Coulomb barrier is ~1 MeV. When two helium nuclei collide, the height of the barrier is ~4 MeV (since each nucleus has two protons). The height of the barrier is ~36 MeV when two carbon nuclei with six protons collide. Let us fix the energy of collision to be the same in all these three cases. Clearly, the *width* of the barrier through which the particles will have to tunnel increases dramatically as the height of the barrier

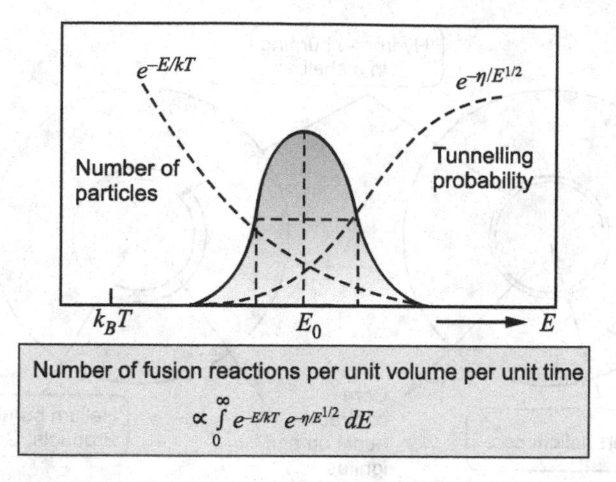

Fig. 11.3 Figure illustrates the significance of the so-called *Gamow peak*. The number of fusion reactions per unit volume per unit time depends upon two factors: *an exponentially decreasing number of particles with increasing energy, and an exponentially increasing tunnelling probability with increasing energy*

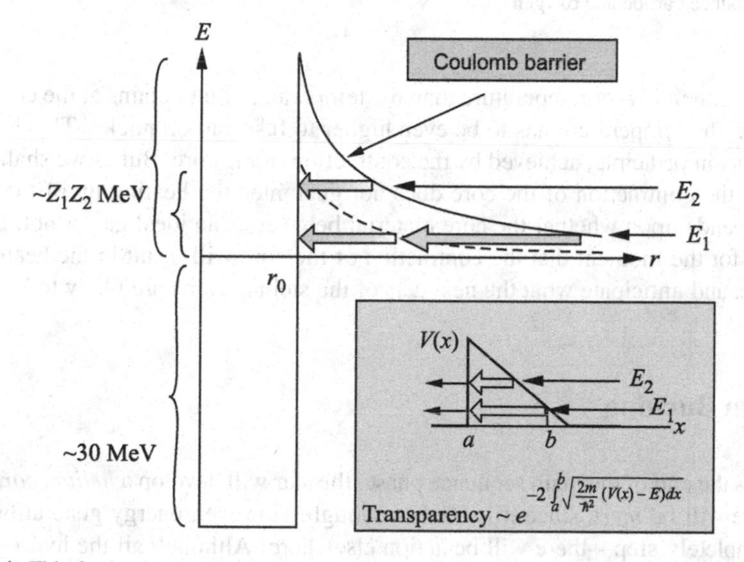

Fig. 11.4 This figure shows why the successive phases of nuclear fusion require higher temperatures. This is primarily because the height of the coulomb barrier increases with increasing electric charge of the fusing particles. For example, the height of the barrier for the fusion of helium is four times higher than for the fusion of protons

increases (refer to Fig. 11.4). And as we mentioned above, the tunnelling probability *decreases exponentially with increasing thickness of the barrier.* It therefore follows that to fuse helium nuclei together the average energy of the particles will have to be much greater than at the centre of the Sun. In other words, the stellar plasma has

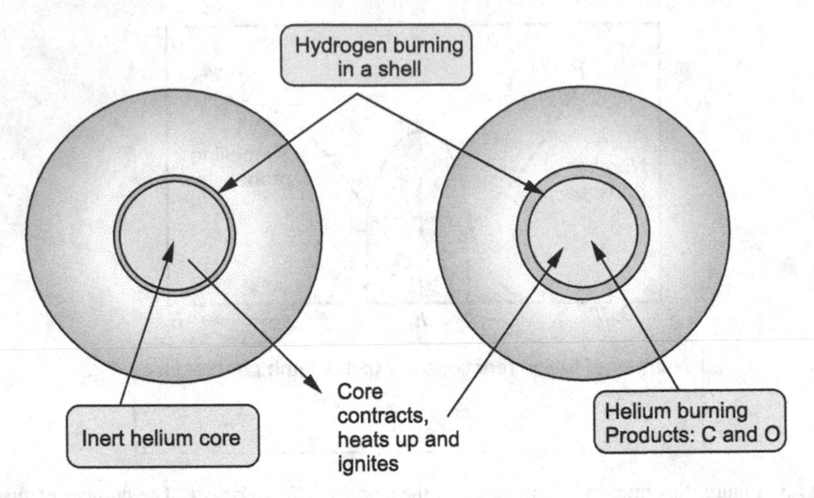

Fig. 11.5 At the end of the main sequence phase, a star will have an inert helium core, surrounded by a shell where hydrogen is still fusing to helium. Since the inert core is not generating heat, it will contract, and consequently heat up. When the temperature reaches about 10^8 K, helium will fuse to produce carbon and oxygen

to be at a much higher temperature than the temperature that obtains at the centre of the Sun. The temperature has to be even higher to fuse carbon nuclei. This heating up can be, in principle, achieved by the contraction of the core. But as we shall soon discuss, the contraction of the core does not guarantee the heating up of the core. This depends upon whether the core material behaves as an ideal gas or not. Let us assume for the moment that the contraction of the core will result in the heating of the core, and anticipate what the next acts of the stellar drama are likely to be.

Helium Burning

Towards the end of the main sequence phase, the star will develop a *helium core.* But this core will be *inert*, since it is not hot enough. However, energy generation will not completely stop—there will be action elsewhere! Although all the hydrogen in the core has been exhausted, there is an enormous amount of hydrogen outside the core. Unfortunately, the temperature is not high enough for hydrogen fusion in the extended envelope of the star. But the hydrogen in a thin *shell* surrounding, and in contact with, the inert core will be hot enough for the synthesis of helium to continue. So the star will have a *shell source* of luminosity, although energy generation in the core has stopped temporarily (see Fig. 11.5).

Since the core is no longer generating heat, the hydrostatic equilibrium would be temporarily disturbed. Gravity would overwhelm the resistance offered by the core and *squeeze it*. As a consequence, the inert helium core will heat up (let us assume that this happens). When the temperature of the core reaches a value of $\sim 10^8$ K, helium nuclei will start to fuse.

The by-products of this phase of fusion will be ^{12}C and ^{16}O. The reactions proceed as follows:

$$\begin{aligned} ^4He + {}^4He &\rightleftarrows {}^8Be, \\ ^8Be + {}^4He &\rightarrow {}^{12}C + \gamma. \end{aligned} \tag{11.5}$$

The key reaction in the fusion of helium is the formation of ^{12}C from three ^4He nuclei. This is known as the *triple alpha reaction* (remember that Rutherford's alpha particle is just a ^4He nucleus). The synthesis of ^{12}C takes place in two steps. In the first step, two alpha particles form a beryllium nucleus. The ground state of ^8Be nucleus is about 100 keV higher in energy and it is therefore unstable. Left to itself, it will decay back to two alpha particles in about 10^{-16} second (this is why we have the arrows pointing in both directions in the first of the two reactions above). Fortunately, the mean interval between the collision of two alpha particles is much shorter than this timescale. Consequently, a third alpha particle can be expected to collide with the beryllium nucleus before it decays back to two alpha particles (the second step in the reaction shown in 11.5). The binding energy released per ^{12}C nucleus formed is about 7.3 MeV. The energy released *per unit mass* in the triple alpha reaction (in which helium is synthesized into carbon) is about 10 times smaller than in the case of the CNO cycle (in which hydrogen is synthesized into helium).

Once a sufficient concentration of ^{12}C has been produced, further capture of alpha particles result in the production of oxygen, neon, etc.:

$$\begin{aligned} ^{12}C + {}^4He &\rightarrow {}^{16}O + \gamma, \\ ^{16}O + {}^4He &\rightarrow {}^{20}Ne + \gamma. \end{aligned} \tag{11.6}$$

Carbon Burning and Oxygen Burning

As a result of helium burning the star will now develop a carbon–oxygen core. But this core will be more centrally condensed and less massive than the original helium core. The reason is that to form the C–O core one needs a much higher temperature $\sim 10^8$ K, and this is likely to obtain only near the centre.

Just as the helium core formed at the end of the main sequence phase was inert, *the carbon–oxygen core will also be inert initially*. For two carbon nuclei to tunnel through the very high Coulomb barrier and fuse, the temperature has to be in excess of 5×10^8 K. The inert carbon–oxygen core will contract and heat up, just as the inert helium core did. When the temperature reaches 500 million degrees, carbon nuclei will fuse. Unlike in the case of hydrogen burning and helium burning, it is a much more complicated business to calculate accurately the end products of carbon burning. The general consensus is that the end products of carbon fusion will be as follows:

$$^{12}\mathrm{C} + {}^{12}\mathrm{C} \rightarrow {}^{20}\mathrm{Ne} \text{ and } {}^{24}\mathrm{Mg}. \tag{11.7}$$

Remember that the core produced by the fusion of helium also contains ^{16}O. Thus, at the end of carbon burning, the core will contain ^{16}O, ^{20}Ne and ^{24}Mg.

Because of the even bigger Coulomb barrier, the fusion of oxygen will not be possible till the central temperature becomes even higher (see Fig. 11.4). When the temperature reaches 10^9 K, oxygen nuclei will fuse to produce a variety of products:

$$\begin{aligned} ^{16}\mathrm{O} + {}^{16}\mathrm{O} &\rightarrow {}^{32}\mathrm{S} + \gamma, \\ &\rightarrow {}^{31}\mathrm{P} + p, \\ &\rightarrow {}^{31}\mathrm{S} + n, \\ &\rightarrow {}^{28}\mathrm{Si} + \alpha \end{aligned}$$

The proton and the alpha particle produced in the above reactions will be immediately absorbed, giving rise to secondary reactions. We shall not go into all that complicated stuff! It would be safe to say that *among the end products of oxygen burning one would find a substantial amount of Silicon* (Si).

Beyond Oxygen Burning

What happens next? It is even more complicated than during oxygen burning. Now the ambient temperature is well in excess of several billion degrees. At these temperatures, some of the more loosely bound nuclei will be broken up. This is known as *Photodisintegration* of nuclei, in analogy with *Photoionization* of the atoms. At $T \geq 10^9$ K the radiation field will consist mainly of gamma rays with energy in the MeV range. These can be absorbed by the nuclei, raising them to *excited states*. Such nuclei in excited states are prone to radioactive decay, emitting alpha particles. As a result of photodisintegration there will be an appreciable number of *free neutrons, protons and alpha particles* inside the core. These will react with silicon and gradually build up heavier elements. This process will go on till ^{56}Fe is reached. If you would like to forget all the intermediate steps, one might loosely say that the next phase after oxygen burning is *silicon burning*,

$$^{28}\mathrm{Si} + {}^{28}\mathrm{Si} \rightarrow {}^{56}\mathrm{Fe}. \tag{11.8}$$

And then the fusion reactions will cease; this is the end of the road. To understand this, we should visit a plot of the binding energy of various nuclei. This is shown in Fig. 11.6.

What is plotted in the figure is the *average binding energy per nucleon* versus *the number of nucleons in the nucleus*. Consider some nucleus of atomic mass number *A*.

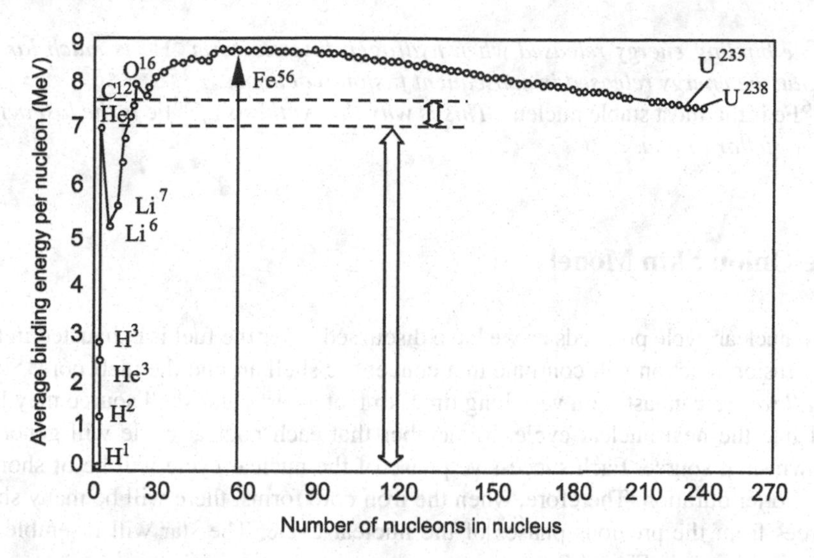

Fig. 11.6 Figure shows the experimentally measured binding energy of nuclei. The y-axis is the average binding energy per nucleon in MeV. The x-axis is the atomic number. As may be seen, ^{56}Fe is the most stable nucleus. Beyond this, the binding energy per nucleon *decreases*. This means that is we want to fuse iron to form heavier elements, the reaction will be endothermic; it will cost us energy

Let its mass be M_{nuc}. A nucleus is stable because its mass is *less* than the sum of the masses of the neutron and protons in the nucleus (recall Aston's discovery). This is known as the *mass deficit*. This mass deficit is equal to the sum of the mass of the $(A - Z)$ neutrons *plus* the mass of Z protons *minus* the mass of the nucleus. The binding energy of the nucleus is equal to (mass deficit $\times$ c^2).

$$E_B = \left[(A - Z)m_n + Zm_p - M_{\text{nuc}}\right] \times c^2. \tag{11.9}$$

When comparing how strongly different nuclei are bound, it is useful to define the *average binding energy per nucleon*,

$$f = \frac{E_B}{A}. \tag{11.10}$$

What is plotted in Fig. 11.6 is the experimentally measured value of this quantity f as a function of A. The important features of this plot that you must observe are the following:

1. f rises sharply from hydrogen, flattens out and reaches a maximum for ^{56}Fe. The maximum value is 8.5 MeV.
2. Beyond ^{56}Fe, the binding energy per nucleon gradually *decreases*.
3. Barring hydrogen, the typical value of f is around 8 MeV.

4. *The binding energy released when hydrogen fuses to form* ^{4}He *is much larger than the energy released in subsequent fusion reactions.*
5. ^{56}Fe is the most stable nucleus. *This is why the synthesis of* ^{56}Fe *is the last act of the stellar drama.*

The Onion Skin Model

If the nuclear cycle proceeds as we have discussed, after the fuel is exhausted in the core, fusion reaction will continue in a concentric shell around the inert core. Such a *shell source* can last for a very long time. In fact a particular shell source may last well into the next nuclear cycle. Remember that each nuclear cycle will generate its own shell source. Each successive phase of the nuclear cycle will be of shorter and shorter duration. Therefore, when the iron core forms, there will be many shell sources from the previous phases of the nuclear cycle. The star will resemble an onion, as shown in Fig. 11.7.

To Burn or Not to Burn

Do we expect every star to develop an onion structure like shown in Fig. 11.7? Do we expect that in all stars the nuclear reactions will go all the way till an iron core forms? We have assumed in the above discussion that the sequence of nuclear cycles proceed without any obstacle. This assumption may or may not be correct. The key assumption we have made is that the inert core will contract and heat up; once the temperature of the contracting core reaches the ignition temperature, the next phase of the fusion reactions will begin. So the crucial thing to settle is whether the contracting core of a star will heat up. And if it heats up, will it get hot enough for the *ashes* of the previous phase to ignite. The answer to this will depend upon the *equation of state*.

Ideal Gas

If the stellar plasma in the core behaves as an ideal gas then it will certainly heat up when the density of the core increases as a consequence of its contraction. It is not difficult to figure out the increase in the temperature for a corresponding increase in the density.

Let us start with the requirement that for the core to be stable the gravitational pressure at the centre must be equal to the gas pressure:

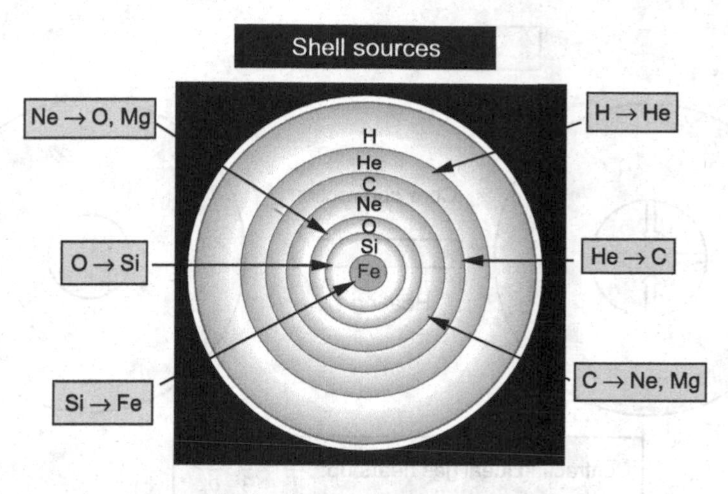

Fig. 11.7 The onion skin model of a star. As we shall see in later chapters, the nuclear cycle will proceed all the way in massive stars. The final stage will be the synthesis of silicon to form an iron core. This will be surrounded by a thin shells consisting of silicon, oxygen, neon, carbon, helium, followed by an extended hydrogen envelope. At this stage, there will be multiple shell sources where fusion reactions are still taking place

$$P_{\mathrm{Grav}} = p_{\mathrm{gas}} \tag{11.11}$$

where

$$P_{\mathrm{Grav}} \sim \frac{GM^2}{R^4}, \; p_{\mathrm{gas}} = \frac{\rho_c k T_c}{\mu m_p}. \tag{11.12}$$

Here ρ_c and T_c are the central density and temperature, respectively (If you are not familiar with the expression for the gravitational pressure, you may refer to the companion volume, ***What Are the Stars***?). Taking the logarithm of both sides in (11.11) and using (11.12) we obtain,

$$\log P_{\mathrm{G}} = \log \rho_c + \log T_c + \text{ constant}. \tag{11.13}$$

The gravitational pressure can be recast as $P_{\mathrm{G}} \propto M^{2/3} \rho^{4/3}$. Taking the logarithm of this we get

$$\log P_{\mathrm{G}} = \frac{4}{3} \log \rho + \text{constant}. \tag{11.14}$$

Since the mass of the star is not a variable, we have absorbed it into the constant in (11.14). Combining (11.14) and (11.13) we obtain the desired relation

$$\boxed{\log T_c = \frac{1}{3} \log \rho_c + \text{ constant}.} \tag{11.15}$$

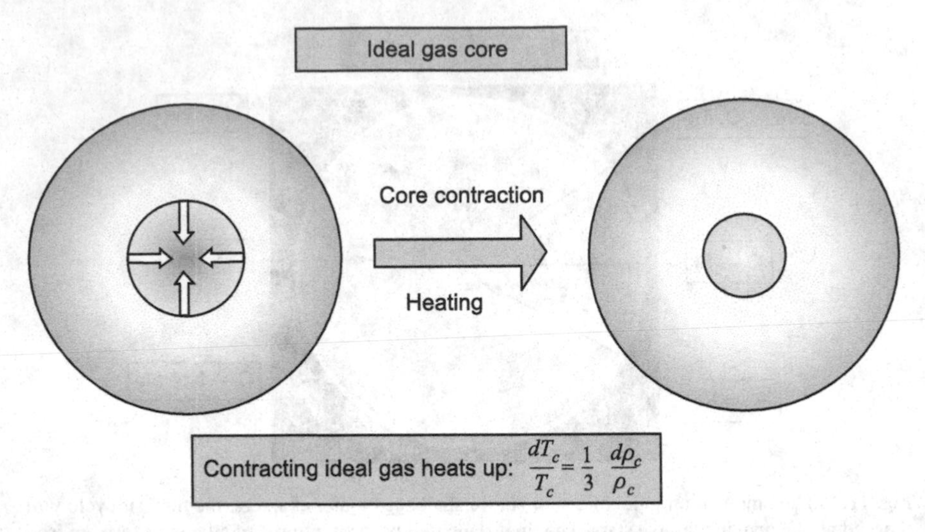

Fig. 11.8 If the core of the star obeys the ideal gas law, then it will heat up upon contraction

The relation (11.15) can also be written as follows:

$$\boxed{\frac{dT_c}{T_c} = \frac{1}{3}\frac{d\rho_c}{\rho_c}.} \tag{11.16}$$

This tells us that *as long as the gas can be regarded as ideal, an increase in the density of the core will result in the core heating up*.

In Eq. (11.11) we ignored radiation pressure. A more general relation would have been

$$\boxed{P_{\mathrm{G}} = \frac{\rho k T}{\mu m_p} + \frac{1}{3} a T^4.} \tag{11.17}$$

The right-hand side of the above equation now includes both gas pressure and radiation pressure. We shall not pause to derive it here, but it can be shown that Eq. (11.16) *is valid even when we include radiation pressure.* The heating up of an ideal gas core upon contraction is shown in Fig. 11.8.

Degenerate Core

If the core is degenerate, however, its response is very different under compression (see Fig. 11.9). Although initially the core might be well described by the equation of state of an ideal gas, there is no guarantee that this would be a good description as its density increases. You may recall from our earlier discussion that the gas

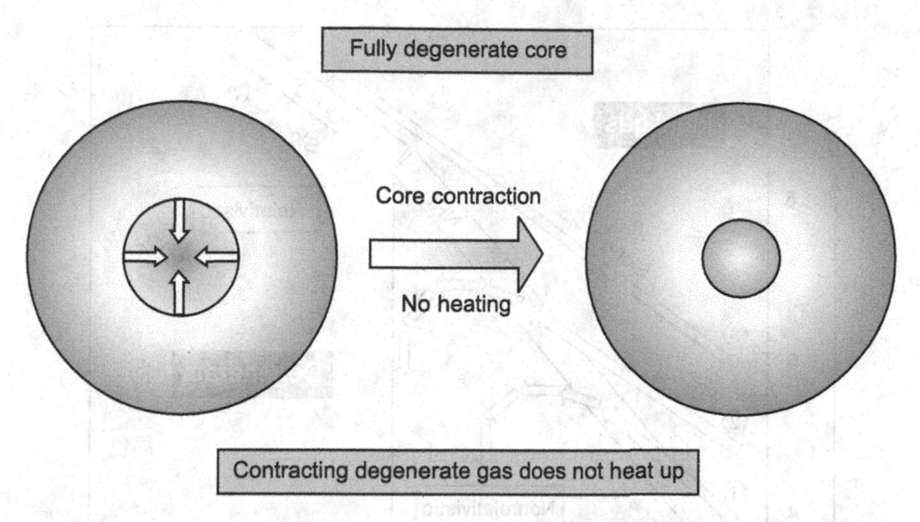

Fig. 11.9 If the core of the star is degenerate, then contraction will not lead to heating. It will just become more degenerate. But, as we shall see, secondary effects, such as a surrounding shell source, can lead to the core heating up

may be regarded as ideal if the thermal energy kT is much greater than the Fermi energy E_F ($kT \gg E_F$). As the density increases the Fermi energy will also increase ($E_F \propto \rho^{2/3}$). At some stage, $kT \approx E_F$, and the gas becomes partially degenerate. As the density increases further, E_F will become greater than the thermal energy kT and the gas should be regarded as fully degenerate ($kT \ll E_F$).

When we compress a fully degenerate gas, its temperature does not increase significantly. The Fermi energy will increase and the gas becomes even more degenerate.

It is instructive to see the behaviour of the gas under compression in a log T versus log ρ plot (see Fig. 11.10). The sloping *rectangular boxes* in Fig. 11.10 represent the transition between the *ideal gas* behaviour (on the left) and *degenerate gas* (on the right). Along the boxes, $kT \approx E_F$. For a *nonrelativistic degenerate gas* $E_F \propto \rho^{(2/3)}$. Therefore, the rectangular box along which $kT \approx E_F$ will have a slope of 2/3 in a log T versus log ρ plot (This box has been labelled as *NRD*). For a *relativistic degenerate gas*, $E_F \propto \rho^{(1/3)}$. Accordingly, the rectangular box labelled *RD* will have a slope of 1/3.

Let us consider the core of a star whose initial position in the diagram is labelled as (1). As the core contracts, it will heat up and its trajectory will be along a path whose slope is 1/3 (see Fig. 11.10). As it contracts even more, it will enter the domain of partial degeneracy. At this stage, a further increase in density will lead to only a marginal increase in temperature. When the core becomes fully degenerate, the density will increase at a *constant temperature*. At some stage, the degeneracy pressure will arrest gravity, and there will be no further contraction. As the gas loses its *fossil heat* its trajectory will swing downward and the gas will cool at a constant density. The core labelled (2) will have a similar trajectory in the plot.

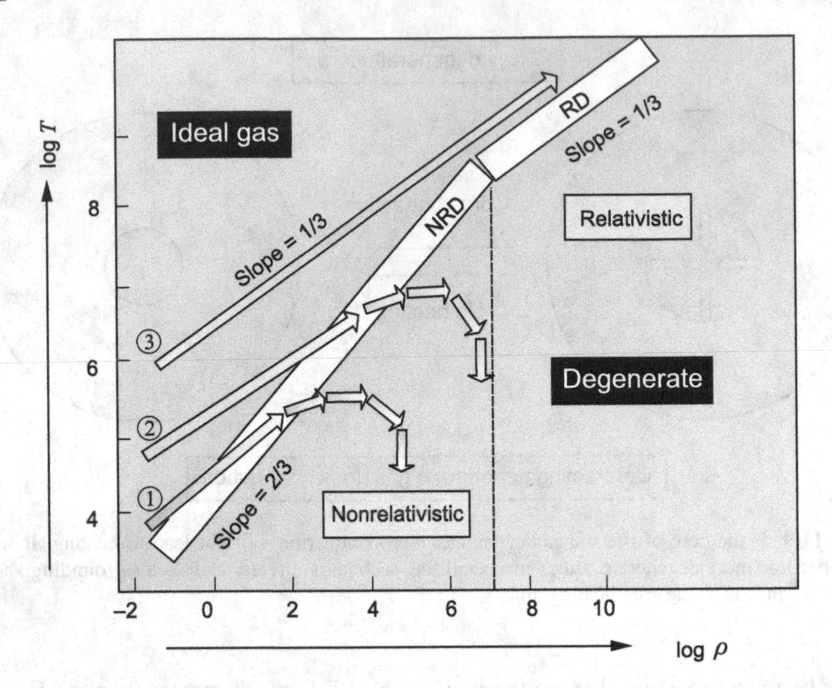

Fig. 11.10 This figure explains the behaviour of an ideal gas sphere under compression. Notice that the temperature and density are plotted in logarithmic units. The 'rectangular boxes' mark the transition from ideal gas behaviour to degeneracy. In the region to the right, the electrons will be degenerate; beyond a density of 10^7 (in cgs units), the electrons will be relativistically degenerate. Along the rectangular boxes, $kT \approx E_F$, which is the criterion for degeneracy to set in. Consider the sphere whose starting point is ②. As we increase the density, the temperature will increase. At some density, which depends upon the temperature, the gas will become degenerate. A further increase in density will not result in an increase in the temperature. At some stage, degeneracy pressure will prevent any further compression. From then onwards, the sphere will cool due to the loss of fossil heat, and it will do so at constant density. In contrast, a sphere whose starting point is ③ will remain nondegenerate at all densities

Now let us consider the core (3). Its initial position corresponds to a much higher temperature. As it contracts it will heat up, and its trajectory will, again, have a slope 1/3. But this time, *the core never crosses the boundary of degeneracy* and it will continue to heat up.

What does all this mean for stars of different mass? This is schematically shown in Fig. 11.11.

Consider two stars of masses M_1 and M_2, and let the initial location of their cores be as shown in Fig. 11.11. Let us first consider the star of mass M_1. As may be seen in the figure, its core is already close to the boundary separating the ideal gas domain and the domain of degeneracy. As the core contracts and the density increases it will heat up. During this process of continued contraction of the core the staris moving to

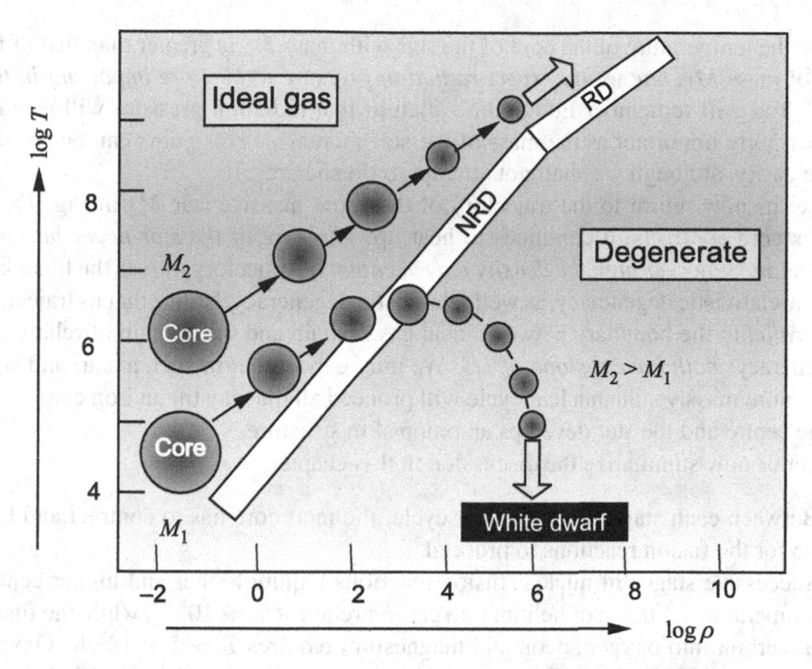

Fig. 11.11 This is essentially the same as Fig. 11.10. Since the fate of star is really decided by its *core*, we have shown the trajectory of the core for two different stellar masses, M_1 and M_2. Since the core of a more massive star will be hotter, for the same density, it is reasonable that $M_2 > M_1$. Simple considerations explained in Fig. 11.10 tell us that stars up to a certain mass will end up as degenerate stars (white dwarfs). *Stars above a critical mass will never become degenerate, however high the density may become*

the right. At some stage degeneracy will set in. After that there will be a brief phase during which the density will increase, with the temperature remaining more or less constant. Soon degeneracy pressure will prevent further contraction of the core and it will become a white dwarf.

What will be the composition of the white dwarf? That depends on the initial mass of the star. *If the central temperature reaches* 10^7 *K, the fusion of hydrogen into helium will take place. If the temperature continues to increase till it reaches* 10^8 *K, then fusion of helium into carbon will occur*. As we shall see in the subsequent chapters, the stellar drama will end here for all stars with initial mass less than about $9M_\odot$. The carbon core will enter the region of degeneracy and will eventually become a C–O white dwarf. Initially the white dwarf will be very hot. But it will cool as the stored heat is radiated away.

Let us stay with Fig. 11.11 and consider the star of mass M_2. The first thing to appreciate is that one expects M_2 to be more massive than M_1 This follows from the fact that although their initial densities are the same, the core of M_2 is hotter. One can therefore say from general considerations that M_2 must be the moremassive star.

Since the temperature of the core of the star with mass M_2 is greater than that of the star of mass M_1, *one would expect radiation pressure to be more important in this core*. You will remember Eddington's dictum that radiation pressure will become increasingly important as the mass of the star increases. This *guess* can be verified quite easily, although we shall not attempt to do so here.

Let us now return to the trajectory of the more massive star M_2 in Fig. 11.11. As its core contracts, it continues to heat up. *The core of the star never becomes degenerate however high the density may become*. Its trajectory misses the boundary of nonrelativistic degeneracy, as well relativistic degeneracy. Notice that its trajectory is *parallel* to the boundary between ideal gas domain and the domain of relativistic degeneracy; *both have a slope of 1/3*. We may expect that in such a star, and stars even more massive, the nuclear cycle will proceed all the way till an iron core forms at the centre and the star develops an onion skin structure.

Let us now summarize the discussion in this chapter.

1. Between each stage of the nuclear cycle, the inert core has to contract and heat up for the fusion reactions to proceed.
2. Successive stages of nuclear fusion reactions require higher and higher central temperatures. Fusion of helium into carbon requires $T \sim 10^8$ K, while the fusion of carbon into oxygen, neon and magnesium requires $T \sim 5 \times 10^8$ K. Oxygen burning requires $T \sim 10^9$ K.
3. The cores produced by successive stages of the nuclear cycle tend to be smaller in mass and more centrally concentrated.
4. If the contracting inert core behaves as a classical ideal gas then it will heat up as it contracts. *The nuclear cycle will proceed uninterrupted as long as this trend of contraction resulting in heating continues, and the necessary ignition temperatures are reached.*
5. However, if degeneracy sets in as the core contracts then it will not heat up further as the core contracts further. Therefore, *the nuclear cycle will be interrupted by the onset of degeneracy.*
6. *Stars up to a certain critical mass will develop degenerate cores and are likely to end their lives as white dwarfs–in most cases, as carbon-oxygen white dwarfs.*
7. *The cores of stars more massive than the critical mass will never become degenerate, however high the density may become*. This confirms the remarkable prediction made by Chandrasekhar in 1932 (refer back to Chap. 7, *The Chandrasekhar Limit*, and Fig. 7.4 in particular).
8. The nuclear cycle will proceed the full course in these massive stars, and the stars will eventually develop an iron core, with many shell sources surrounding it.

This is as far as one can go with the kind of qualitative reasoning we have been pursuing. To go beyond this, one has to resort to actual numerical calculations with fast computers. There is one other thing: so far, we have concentrated exclusively on what is happening in the core of the star. Admittedly, this is where the energy generation is taking place. But how does the rest of the star respond to all this? After all, what we see is the surface of the star!

In the next four chapters we shall briefly summarize what modern investigations have taught us about the life history of stars. Since the life history depends upon the mass, we shall break up the discussion into three parts: low mass stars like the Sun, intermediate mass stars and massive stars. In the next chapter, we shall outline the life history of the Sun.

Chapter 12
What Does the Future Hold for the Sun?

Early Evolution

In the preceding chapter we discussed the evolution of the core of the star. We saw that the life history of the *core* depends upon the mass of the star. But the core is not what we see! We shall now turn to a brief discussion of what modern calculations tell us about evolution of the stars, in particular, how the outer layers of the star respond to the evolution of the core. The evolution of stars is best understood by looking at the *Hertzsprung–Russell diagram* we introduce in Fig. 12.1. The movement of a star in this diagram, as it evolves, will tell us how its radius, temperature and luminosity vary in response to the behaviour of the core of the star. As we shall see, the behaviour of the stars in the lower part of the *Main Sequence* is in some respects qualitatively different from that of the more massive stars. Therefore, we shall divide our discussion into three parts, as indicated in Fig. 12.1. In this chapter, we shall discuss stars with mass less than about $2.5M_\odot$. Specifically, we shall consider how our Sun will evolve in the future. In the next chapter we shall discuss stars in the *intermediate mass range*, that is, stars in the mass range $2.5M_\odot$–$9M_\odot$. Finally, we shall take up the evolution of stars more massive than, say, $10M_\odot$.

If you refer back to Chap. 2, Fig. 2.6, you will notice that an essential difference between stars in the lower and upper part of the main sequence is that low-mass stars have *radiative cores*, while intermediate-mass stars and massive stars have *convective cores*. This has a major implication for how the helium core grows in mass as the star consumes the hydrogen in the central region. In the Sun, for example, the helium core grows in mass rather *slowly*, starting with zero mass. The helium core, which is the result of fusion of hydrogen, will be inert. This is because the temperature of the helium core will be much less than 10^8 K needed for helium to fuse. In fact, since there is no energy generation in the inert helium core, its temperature will be essentially that of the *shell* surrounding it in which hydrogen is being fused into helium. The helium produced in the *shell* adds to the mass of the core. And this process of the growth of the mass of the core is rather slow since the efficiency of helium production in the p–p chain reaction is rather small.

G. Srinivasan, *Life and Death of the Stars*, Undergraduate Lecture Notes in Physics,
DOI: 10.1007/978-3-642-45384-7_12, © Springer-Verlag Berlin Heidelberg 2014

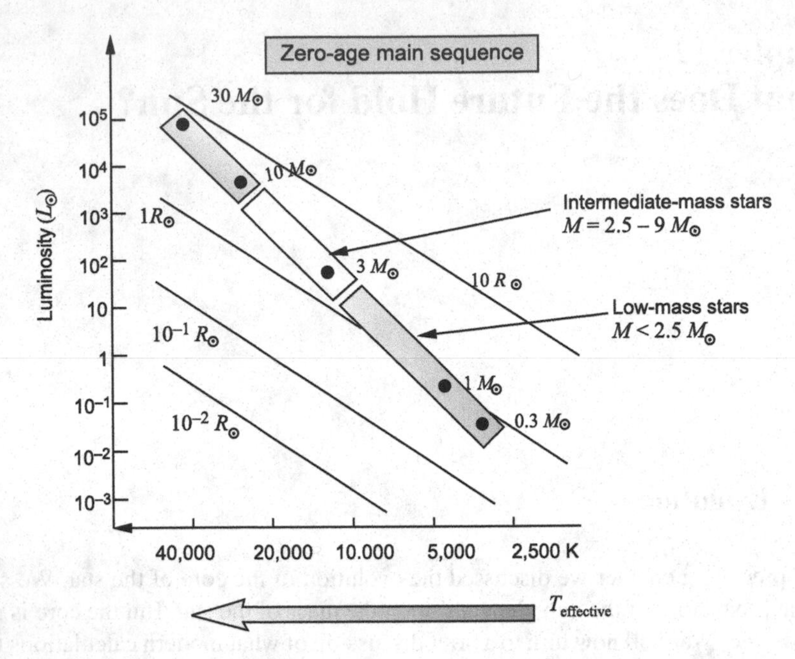

Fig. 12.1 The Hertzsprung–Russell diagram showing the main sequence of stars. In this chapter, we shall discuss the life history of stars in the lower part of the main sequence, namely, stars less massive than 2.5 $M_\odot$

Therefore, this phase in which there is an inert helium core, surrounded by a shell in which hydrogen is burning, lasts for a very long time—*a nuclear timescale*. Consequently, one may expect to see many stars in the sky which are in this phase. Two important things happen to the star during this phase:

1. The envelope of the star will expand dramatically, and
2. The luminosity of the star will increase nearly a thousandfold.

Let us discuss these two points in some detail.

The Star Becomes a Red Giant

Since the helium core is *inert* it will contract due to the weight of the overlying layers of the star. As the core contracts, the gravitational energy released will lead to an expansion of the outer layers. This is an example of a general behaviour known as *gravo-thermal catastrophe,* illustrated in Fig. 12.2. The basic point is that self-gravitating systems have *negative specific heat* (we have encountered this before). *If heat is allowed to flow between two such systems, the hotter one loses heat and gets hotter while the colder system gains heat and gets even colder!* In the case of the star we are considering, a contraction of the core will lead to an expansion of the outer layers.

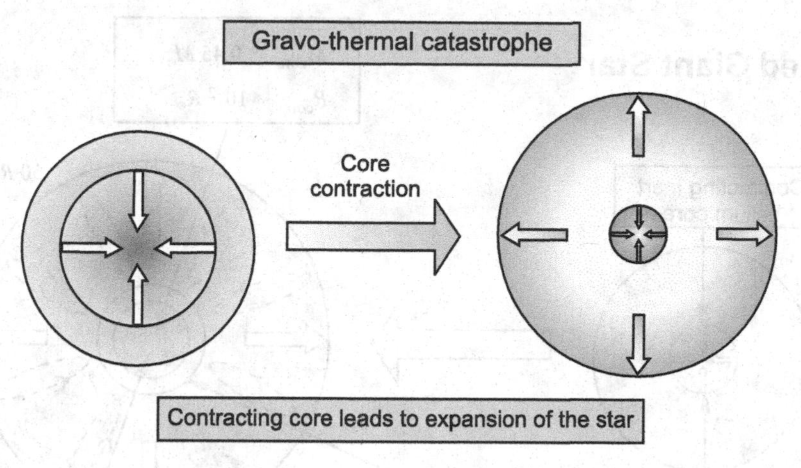

Fig. 12.2 This figure illustrates a general principle, namely, when the core of the star contracts, the envelope will expand. The energy for lifting the envelope comes from the gravitational energy gained in the contraction of the core

Such a behaviour is quite generic. It is important in stellar systems known as *Globular Clusters*. These are gravitationally bound systems of nearly a million stars; with each star moving in the average gravitational potential of all other stars, just as atoms in a star are held together by the combined gravity of all other atoms. These globular clusters undergo what is known as *core-collapse*—the *core* of the star cluster suddenly contracts. When this happens, the cluster as a whole expands.

As the star expands, its surface temperature will decrease and it will become a *red star*. Thus, our Sun will become a red giant even as its inert helium core contracts.

The Ultraluminous Giant Star

As the inert helium core grows in mass, the luminosity of the hydrogen burning shell surrounding it will increase. This may be traced to a very simple reason. As the core contracts, the *surface gravity* of the core will increase (this is just another terminology for *acceleration due to gravity at the surface*). This enhanced gravity will squeeze the thin shell of hydrogen that is burning outside the core, increase its density and temperature, and thus increasing the luminosity of the shell source. Detailed considerations show that in this prolonged phase during which the entire luminosity of the star is generated by the shell source, the luminosity is essentially determined by the mass and radius of the core, and is independent of the mass of the star. And the luminosity is a *strong function* of the mass of the core:

$$L \sim M_{\mathrm{core}}^{7} \qquad (12.1)$$

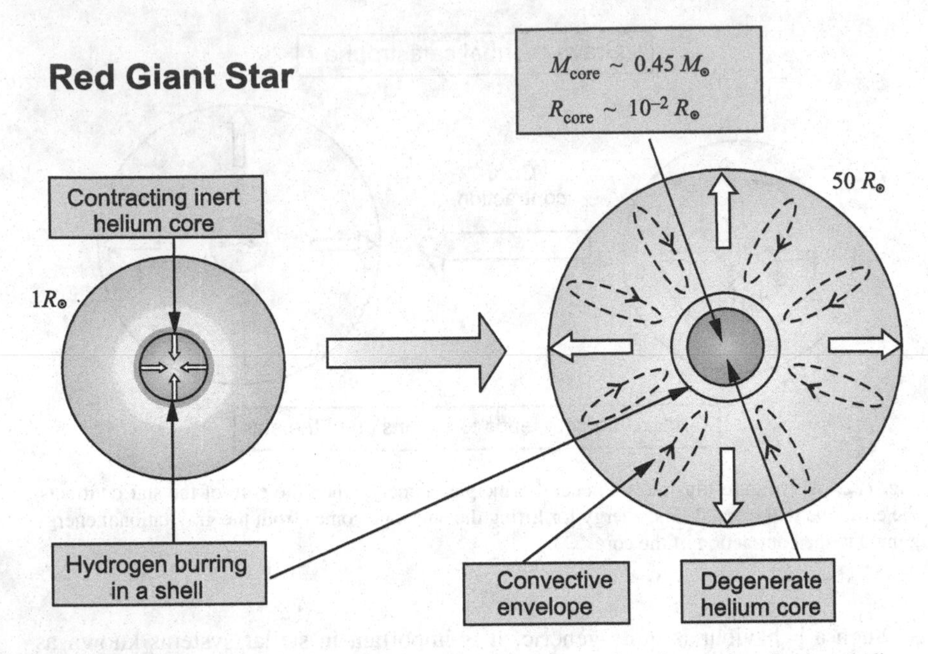

Fig. 12.3 The structure of a red giant star. *Right* at the *centre*, there is a contracting inert helium core. The luminosity of the star is generated in a hydrogen burning shell surrounding the inert core. The envelope expands as the core contracts, and the star becomes a *giant*. Simultaneously, the envelope of the star becomes convectively unstable

Detailed calculations show that the luminosity of the star will increase by a factor of thousand during this phase.

It is important to bear in mind that as the core contracts, it will become *degenerate*. Remember that the central density of the Sun is already ~150g.cm^{-3}. So the core is very close to becoming degenerate. When the central density increases further, the helium core will become degenerate (refer back to Chap. 11, Figs. 11.10 and 11.11). *As the mass of the degenerate helium core grows, its radius will decrease.* (Recall the *inverse* relation between the mass and radius of white dwarfs.) The degenerate core is not quite the white dwarf we discussed in earlier chapters, but it is getting there!

The internal structure of a red giant star is shown in Fig. 12.3. An important transformation has occurred to our star as it expanded to become a giant star—*it has become fully convective*. We saw in Fig. 2.6 that stars in the lower part of the main sequence have convective envelopes whereas a star in the upper main sequence would have a radiative envelope. And we discussed plausible reasons for this important difference. To understand the convective outer layers of the Sun, we invoked an increase in the opacity of the outer layers due to new species of absorbing ions, such as the *negative ion of hydrogen*. But there is a more fundamental, and *generic*, reason for the convective envelopes. And that has to do with their location in the H–R diagram. We shall now mention this in passing. Let us look at Fig. 12.4.

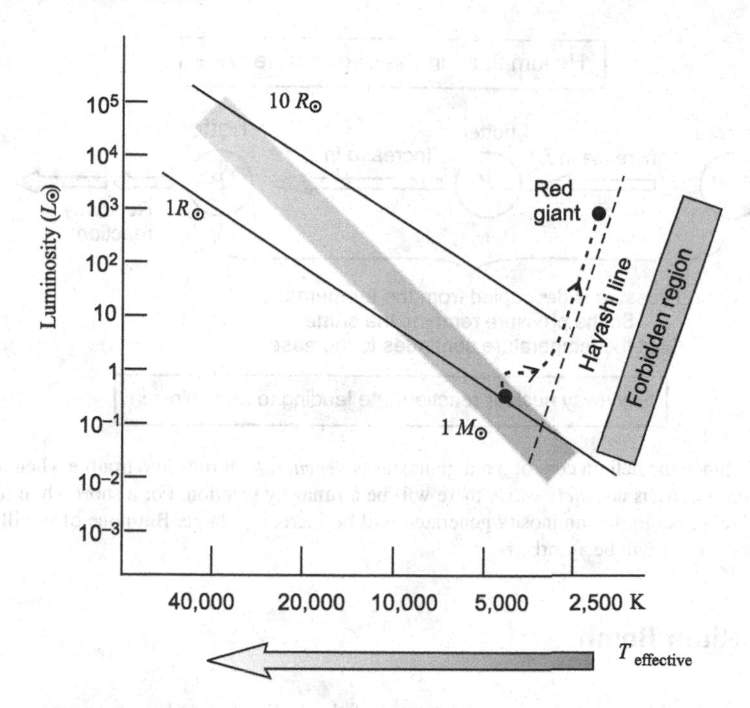

Fig. 12.4 *The Hayashi line.* The nearly *vertical dashed line* is the boundary between stable stars (to the *left*) and unstable stars (to the *right*). Stars along the Hayashi line will be fully convective. As the Sun evolves out of the main sequence, it will move to the *right* in the H–R diagram. As its radius increases, and the surface temperature decreases, it will encounter the Hayashi line. Since it cannot cross the line, it moves vertically and becomes a giant star

We have already commented that as the core of the star contracts its envelope will expand; initially, the luminosity generated by the hydrogen burning shell remains more or less constant. This will make the star move towards the right in the diagram (look at the lines of constant radius in Fig. 12.4). Since its *surface* temperature will decrease as moves towards the right, the temperature gradient in the star will increase. At some stage, the temperature gradient will exceed the *critical* or *adiabatic temperature gradient* and the star will become convective. The near-vertical dashed line labelled the *Hayashi line* in Fig. 12.4 is the locus of fully convective stars (strictly speaking, it is the locus for a star of given mass and chemical composition). That means that all stars on that line will be fully convective. But the Hayashi line is far more significant. Without going into a detailed argument we shall merely mention that *the region to the right of the Hayashi line is forbidden for stars in hydrostatic equilibrium.* The track of the star as it evolves to a red giant is shown in the figure. As the star moves to the right of the diagram, it encounters the Hayashi line. At that point, it becomes fully convective. Since it cannot cross this line, it moves up along it as its luminosity increases dramatically.

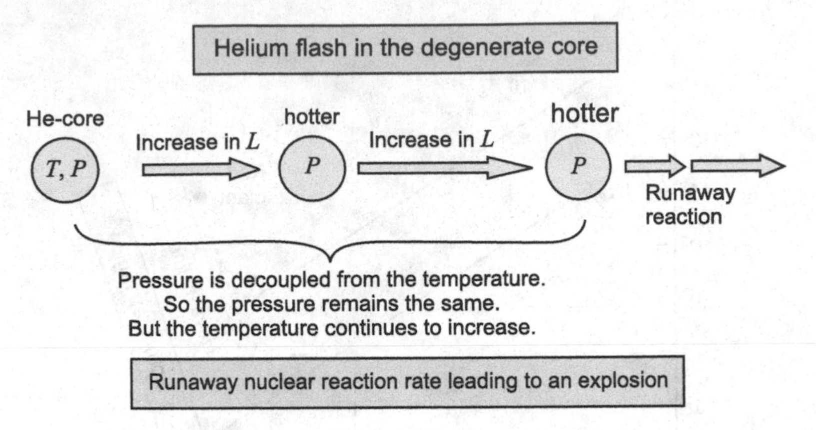

Fig. 12.5 Since the helium core of a red giant star is *degenerate*, it runs into trouble when helium ignites. Since there is no *safety valve,* there will be a runaway reaction. For a short while (till the safety valve turns on) the luminosity generated will be incredibly large. But none of it will reach the surface since it will be absorbed

The Helium Bomb

Even as the star becomes a giant, the mass of the helium core has been growing. As the mass of the core grows, and it contracts, its temperature will also increase. This may seem contrary to what I had said earlier. I had argued that whereas an ideal gas core will heat up upon contraction, a contracting degenerate core will not heat up (see Figs. 11.8 and 11.9). That is true. In the present context, the temperature of the core increases due to a secondary reason. Because the degenerate core will have very high thermal conductivity, it will be at the same temperature as the surrounding shell in which hydrogen continues to burn. As mentioned earlier, as the core mass increases the shell luminosity increases dramatically, as in Eq. (12.1), and the temperature of the shell increases. If you like, *the core is surrounded by a hot plate which is getting hotter and hotter*. At some stage, the temperature of the core will reach 10^8 K and helium can ignite. *Numerical calculations show that helium begins to fuse to carbon and oxygen when the mass of the helium core grows to*

$$\boxed{M_{\text{core}}(\text{at helium ignition}) \approx 0.45 M_{\odot}.} \tag{12.2}$$

Remarkably, this critical mass of the core when helium can ignite is independent of the mass of the star. The star is so distended by now ($R \approx 50 R_{\odot}$) that the core is not affected by the overlying envelope of the star!

When helium ignites in the degenerate core, all hell breaks loose! The reason can be seen in Fig. 12.5.

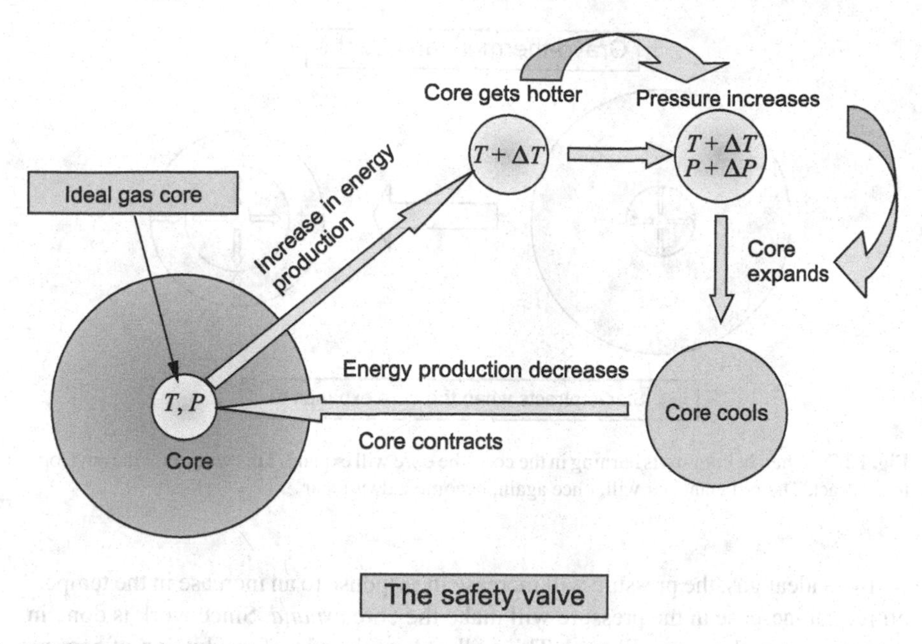

Fig. 12.6 As the degenerate helium core gets hotter and hotter—due to the runaway energy generation—a stage will be reached when the core is no longer degenerate; kT will become much greater than E_F. Once the core ceases to be degenerate, an increase in the temperature will lead to an increase in the pressure, and the safety valve will turn on. The star has been saved from a catastrophe!

When helium ignites in the core, the core which was inert till now has a new source of energy. This will increase the temperature of the core and the fusion reaction will go faster. But because the core is degenerate, the pressure of the core will not increase in response to an increase in the temperature. Remember that when $kT \ll E_F$ (which is the condition for strong degeneracy) the pressure is, to a very good approximation, independent of temperature; the pressure is $\propto \rho^{\frac{5}{3}}$. Since the pressure does not increase, the core cannot expand. Since the core is hotter now than when helium fusion started, the reaction will proceed even faster. The triple alpha reaction that we discussed in Chap. 11 is extremely sensitive to the temperature, with the reaction rate increasing as T^{40}. Therefore we will have a runaway energy generation—a helium bomb! *For a few seconds after helium ignites in the core, the luminosity generated will be* $\sim 10^{11}$ *solar luminosity*! This enormous amount of energy is easily absorbed by the star.

Why does the star not blow up? It is because the helium bomb will soon fizzle out! As the temperature of the core increases dramatically, the core will become less and less degenerate. Soon, kT will become comparable to E_F, and then $kT \gg E_F$. The core is no more degenerate; it will be an ideal gas. And when this happens, the safety valve comes into operation. This is described in Fig. 12.6.

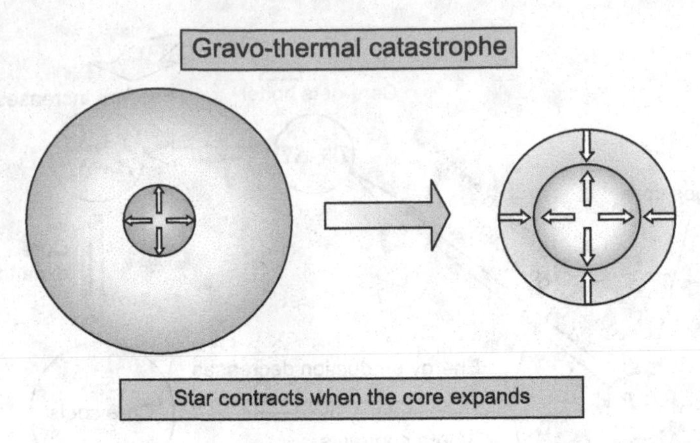

Fig. 12.7 When helium starts burning in the core, the core will expand. This will cause the envelope to contract. The red giant star will, once again, become a dwarf star

In an ideal gas, the pressure will increase in response to an increase in the temperature. An increase in the pressure will make the core *expand.* Since work is done in the expansion, the core will cool. This will reduce the rate of production of energy, causing the core to once again contract to the original radius. This is the safety valve that is at work in the Sun. It is this safety valve that prevents the Sun from blowing up.

To summarize the above discussion, when helium ignites in the core of the Sun—some 7 billion years from now—it will very nearly blow itself up. Fortunately, the safety valve will come into operation before this happens. But, as we shall see in the next chapter, such a rescue is not guaranteed at later stages of evolution!

Helium Burning in the Core

Once the core becomes *nondegenerate,* and the imminent danger has passed, helium will begin to fuse in a steady manner in the core. Remember that hydrogen is still burning in a shell surrounding the core. The onset of energy production in the core will cause the core to expand. And when the core expands, the star will contract; the *mirror principle* once again in operation (see Fig. 12.7). This will cause the star to contract. *The expansion of the core has another consequence. The temperature of the core, as well as the hydrogen burning shell, will decrease, resulting in a decrease in the luminosity of the star compared to the red giant phase.* Consequently, the Sun will descend in the H–R diagram.

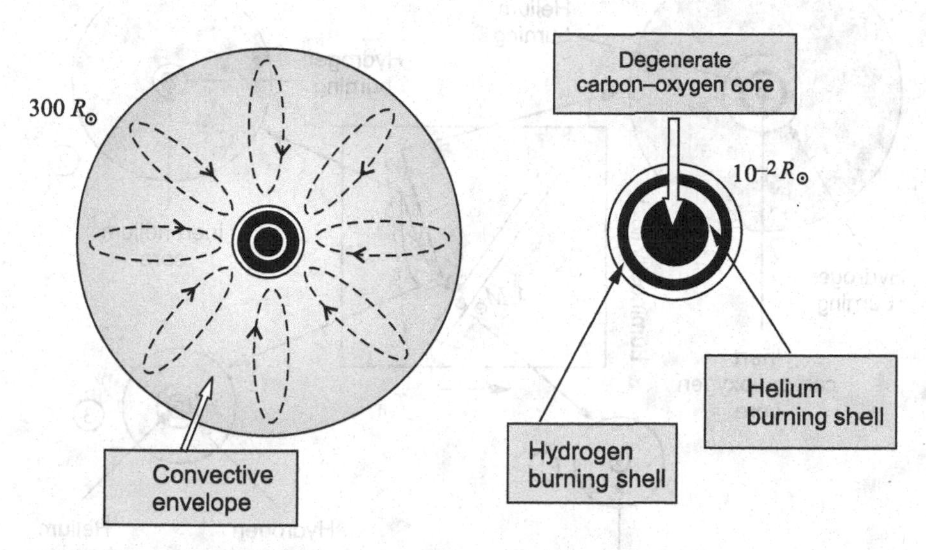

Fig. 12.8 The Red Supergiant star. The contracting inert carbon–oxygen core causes the star to expand to become a supergiant. When the Sun becomes a supergiant, its radius will be 300 times larger than the present radius—much bigger than the radius of Earth's orbit around the Sun. The Earth and the inner planets will be engulfed by the Sun! The inert degenerate core will be surrounded by a helium burning shell, and also a hydrogen burning shell

The Supergiant Star

Two things will happen in this phase. The mass of the helium core will grow due to hydrogen being fused to helium in the shell surrounding the core. At the same time, helium will be converted to carbon and oxygen (as discussed in the previous chapter). After some time a *carbon–oxygen core* will be formed at the centre of the helium core. The inert carbon–oxygen core will be surrounded by *two shell sources;* helium will be fusing in the inner shell, and hydrogen will be fusing in the outer shell.

The inert carbon–oxygen core will contract due to the weight of the overlying layers, just as the inert helium core had contracted earlier. As a consequence, the star will expand once again. Since the carbon–oxygen core is much more compact and dense than the original inert helium core, the gravitational potential energy released during its contraction will be much greater. The star will thus expand to an even larger radius than the red giant, and become a *red supergiant star.* The internal structure of such a red supergiant star is shown in Fig. 12.8.

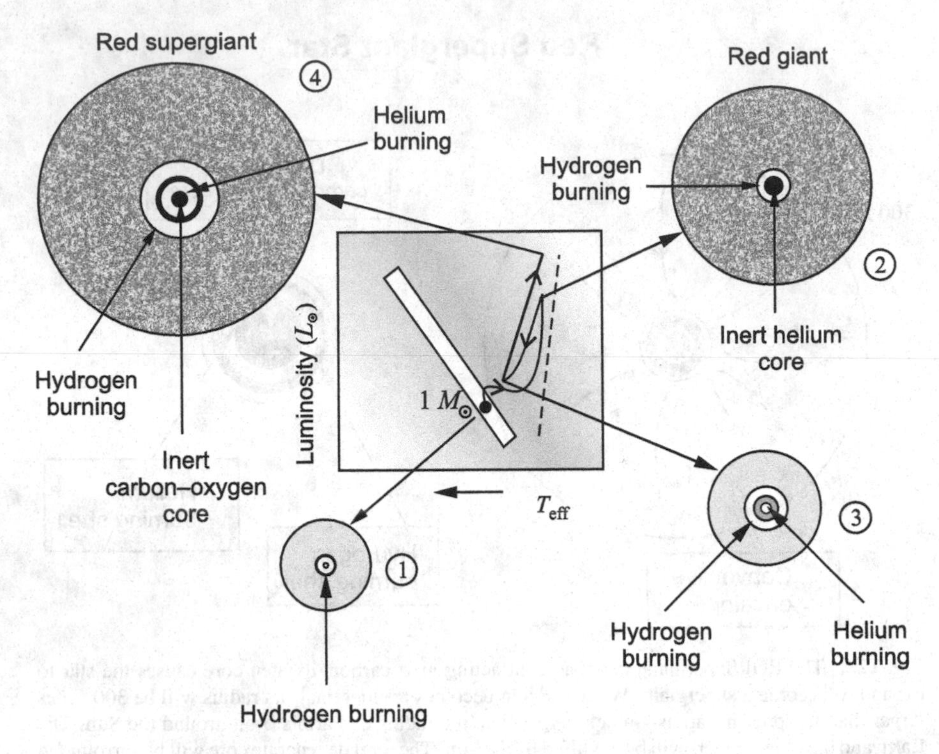

Fig. 12.9 The panel at the centre of the figure shows the trajectory of a star like the Sun in the H–R diagram. The structure of the star in the various phases is also shown

Notice that the radius of the supergiant is about 300 solar radii, which means that when the Sun becomes a supergiant it will engulf the Earth and the other inner planets. But do not worry; it is not going to happen for several billion years!

Figure 12.9 summarizes the evolution of the Sun from its present phase on the main sequence till it becomes a red supergiant.

The Observational Hertzsprung-Russell Diagram

Let us now confront these theoretical results with observational data. The best way to do this would be to construct an H–R diagram with actual data and compare it with the inset at the centre of Fig. 12.9. Although such a H–R diagram has been constructed using observations that stretched over a century, the main difficulty was an accurate determination of the distances to the stars. Obviously, this distance would be needed in the calculations to convert the observed brightness of the stars to their intrinsic brightness or luminosity. Parallax measurements would be one of the best ways to determine the distance to nearby stars. You will recall that *parallax* is the

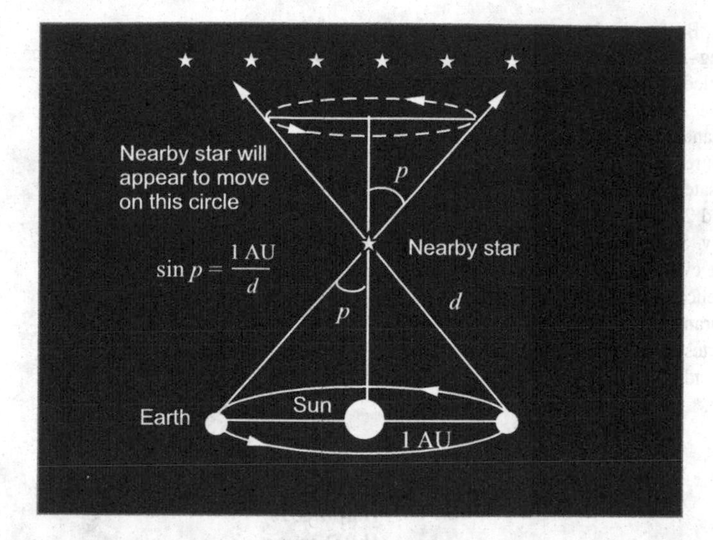

Fig. 12.10 This figure explains the concept of stellar parallax. The line of sight to a nearby star with respect to very distant stars will change as the Earth moves in its orbit around the Sun. Since we know the radius of the orbit, one can estimate the distance to a nearby star by measuring the parallax angle defined in the figure

angular shift in the position of a star in the sky with respect to very distance stars, as the Earth moves around the Sun. This is explained in Fig. 12.10.

You are undoubtedly familiar with the concept of *parallax.* Imagine you are travelling in a train, and have a window-seat. The nearby scenery (the trees, the lamp posts, or houses) will appear to *move* with respect to a distant hill or some such land mark. This is what is known as *parallax shift*. You will see a similar phenomenon when you look at the stars in the sky. *The line of sight to a nearby star with respect to very distant stars will change as the Earth moves in its orbit around the Sun.* Every six months, the Earth moves nearly 300 million kilometres. The Earth–Sun distance (defined as *one Astronomical Unit*) is roughly 150 million kilometres (or 150 solar radii). Therefore, the diameter of Earth's orbit around the Sun is roughly 300 million kilometres. Obviously, the parallax shift would be maximum when the Earth is at diametrically opposite points in its orbit. As the Earth revolves around the Sun, the nearby stars will appear to move on a circle with respect to the distant stars. The relation between the parallax shift and the distance to a nearby star is explained in Fig. 12.10. You may be interested in a little bit of history concerning stellar parallax.

The ancient Greeks debated whether the Universe was *Heliocentric* or *Geocentric*. One of the staunch exponents of the Heliocentric system was *Aristarchus* of Samos (320 BC–250 BC). He explicitly stated that if the Earth revolved around the Sun, then the nearby stars would appear to move in the sky with respect to the distant stars. He looked for this effect but could not detect any parallactic motion. But he did not abandon his heliocentric model of the Universe. Instead, he concluded that the stars must be very far away! A little later, the very well known Astronomer, mathematician

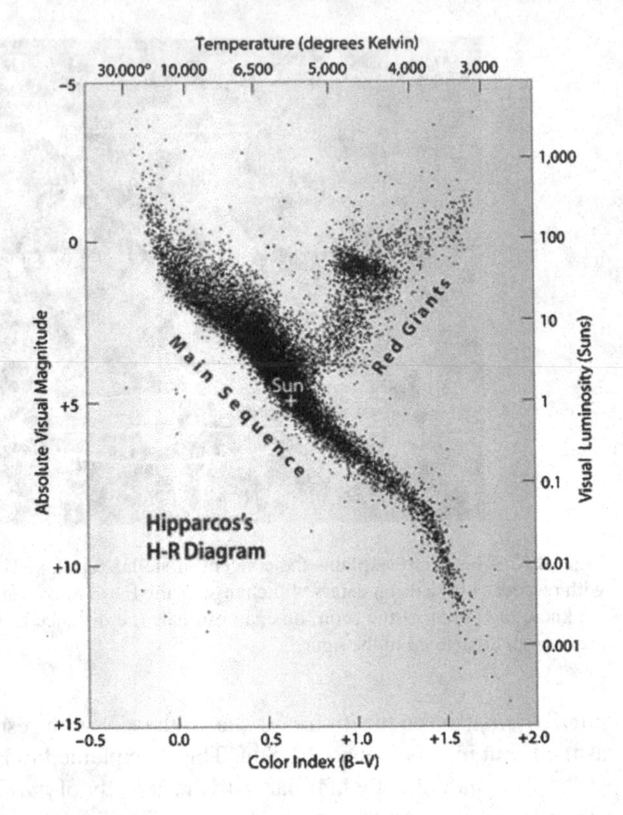

Fig. 12.11 By far the best Hertzsprung–Russell diagram ever compiled. This diagram includes the 20,853 stars whose distances and colours were measured by the HIPPARCOS satellite with better than 10 and 25 % accuracy, respectively. Sunlike stars which have evolved out of the main sequence, and ascending the giant branch, are clearly seen [Courtesy of Michael Perryman and ESA]

and geographer *Hipparchus* (190 BC–120 BC) undertook a systematic study to detect the parallactic motion. Incidentally, Hipparchus was the founder of Trigonometry! He, too, could not detect any parallax. The parallax shift in the position of the stars is very tiny (*less than one second of arc*) had to wait for the advent of the telescope to detect this. The famous astronomer and mathematician *Bessel* had the privilege of detecting the first parallax shift in 1838.

In 1989, the European Space Agency launched a satellite named **HIPPARCOS** (High Precision Parallax Collecting Satellite) to accurately measure the distances to stars—the acronym was chosen in honour of the prescient efforts by Hipparchus, more than two thousand years earlier! HIPPARCOS has accurately measured the distance to more than a million stars. One of the things this has enabled astronomers to do is to construct the Hertzsprung–Russell diagram with a large number of stars. Figure 12.11 is the H–R diagram constructed with the data obtained with the HIPPARCOS satellite. This diagram contains 23,000 stars! The Main Sequence is clearly seen. We also see a large number of sunlike stars ascending the Giant Branch. Several white dwarfs are also seen in the bottom left-hand side corner of the diagram.

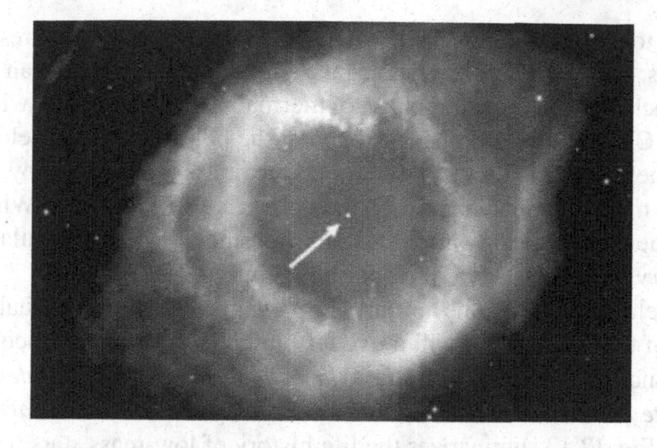

Fig. 12.12 *The Helix Nebula.* At the centre of the envelope ejected by a red supergiant one can see the core of the star. When this cools, it becomes a degenerate white dwarf. Such nebulae are known as *Planetary Nebulae.* [Courtesy of NASA, ESA, C.R. O'Dell (Vanderbilt University), and M. Meixner, P. McCullough, and G. Bacon (Space Telescope Science Institute)]

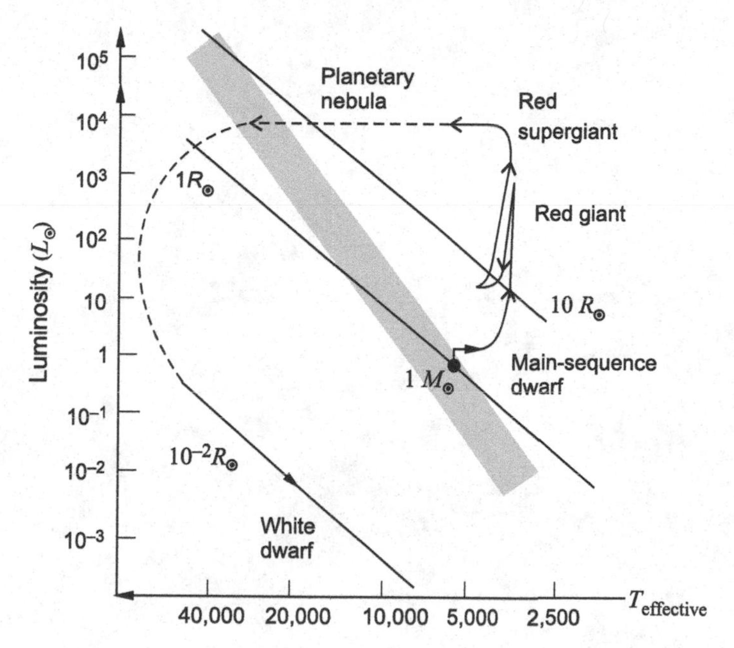

Fig. 12.13 A summary of the life history of a low-mass star like the Sun

Thermal Pulses and Mass Ejection

When the Sun becomes a supergiant, it will find itself in a precarious situation. The two shell sources in the interior become *coupled* and *thermally unstable.*

This leads to a cyclic phenomenon known as *thermal pulses*. As a result of these thermal pulses, the luminosity and the surface temperature of the star can vary appreciably with each pulse. During this phase of thermal instability the star will lose mass dramatically. On the one hand, the hydrogen at the bottom of the envelope is being converted to helium in the shell. At the same time, mass is being lost by the surface. While all this makes sense qualitatively, the details are still not clear. What is amply clear is that the mass ejection during this phase results in what is popularly referred to as *Planetary Nebulae*, such as the one shown in Fig. 12.12.

As the envelope is ejected, the hot carbon–oxygen core will be gradually exposed. What is left of the original star will move left in the H–R diagram since the surface temperature increases towards the left. *When the envelope is completely lost, only the degenerate core remains and the Sun will at last find peace as a carbon–oxygen white dwarf!* Fig. 12.13 summarizes the life history of low-mass stars like the Sun.

Chapter 13
Life History of Intermediate Mass Stars

The Helium Core

We shall now discuss the evolution of stars in the intermediate mass range of 2.5 to 9 $M\odot$. These are located in the upper part of the main sequence (see Fig. 12.1). An essential difference between these stars and low-mass stars like the Sun is the nature of the helium core and its behaviour. In the Sun, for example, the core is in *radiative equilibrium*. The formation of the helium core (due to the fusion of hydrogen) depends only on the *local* rate of helium production. This results in the gradual formation of the helium core, starting with zero mass. And once formed, the helium core is in a degenerate state. The mass of the core grows essentially due to hydrogen burning in the shell surrounding the core. Thus the growth of the core occurs over a *nuclear timescale*, lasting many billions of years. This is why the transformation of the star into a giant is a gradual process, and we can catch the stars in this phase.

In contrast, the cores of stars in the upper part of the main sequence are *convective*. Because of convection, more and more hydrogen from the outer periphery of the central region is brought into the central region, where conditions are right for fusion reactions. This hydrogen is, in turn, converted to helium. *Thus one expects a substantial helium core at the end of the central hydrogen burning phase*. And this *helium core will not be degenerate*.

The Schönberg–Chandrasekhar Limit

Thus, at the end of the main sequence phase, a star of intermediate mass will have a well formed helium core, surrounded by a hydrogen-rich envelope. Since the helium core will be inert, it will be isothermal. Models of such stars—with isothermal helium core surrounded by a hydrogen envelope—were studied by Chandrasekhar and his Research Associate by name Schönberg in 1942. Their most important conclusion—as borne out by subsequent developments—was the following:

G. Srinivasan, *Life and Death of the Stars*, Undergraduate Lecture Notes in Physics,
DOI: 10.1007/978-3-642-45384-7_13,

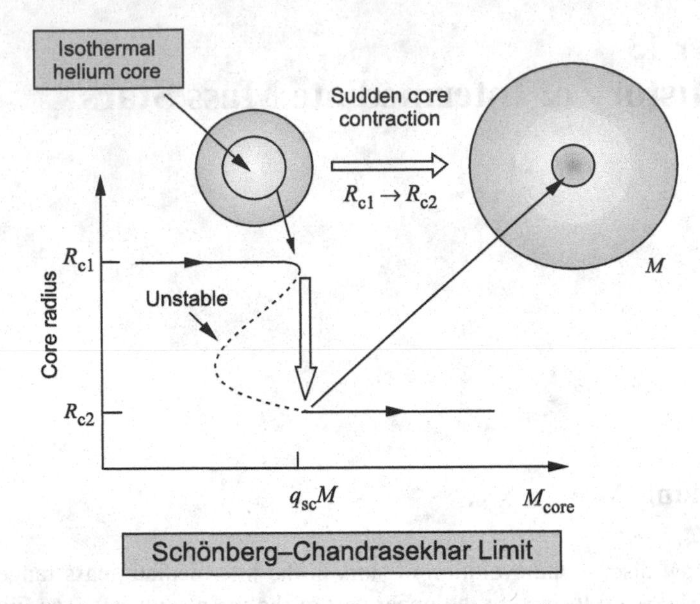

Fig. 13.1 This is a plot of the radius of a *nondegenerate isothermal core* versus the mass of the core. As the core grows in mass, the star moves along the upper branch. When the mass of the core reaches a critical value—roughly 10 % of the mass of the star—then the only stable branch corresponds to a much smaller core radius. Therefore, the core suddenly contracts. This results in a sudden expansion of the star into a giant

- *There are no equilibrium configurations with the isothermal cores having mass exceeding a critical mass.*

Schönberg and Chandrasekhar estimated this upper limit to the mass of the isothermal core to be roughly *ten percent of the mass of the star.*

$$\boxed{M_{\text{core}}\ (\text{Schönberg–Chandrasekharlimit}) \sim 0.1 M_{\text{star}}.}$$

It is customary to define $q = M_c/M$ as the ratio of the mass of the core to the mass of the star. The upper limit to the mass of the isothermal core has come to be known as the Schönberg–Chandrasekhar Limit: $q_{SC} = 0.1$. This limit is certainly exceeded by the helium cores left behind after central hydrogen burning in stars of the upper main sequence. What, then, is the significance of this limit?

This is explained in Fig. 13.1. What is shown in this figure is a series of equilibrium solutions—with an isothermal core and an envelope—for a star of 3 solar mass. Plotted along the y-axis is the radius of the isothermal core and the mass of the core is plotted along the x-axis. As will be seen, there are *three branches* to the curve. The solid lines represent thermally stable branches and the *dashed section represents thermally unstable models*. When the mass of the core is small, the star is in the upper branch; the core is nondegenerate along this branch. This corresponds to a dwarf star close to the main sequence. As the mass of the core increases due to hydrogen burning

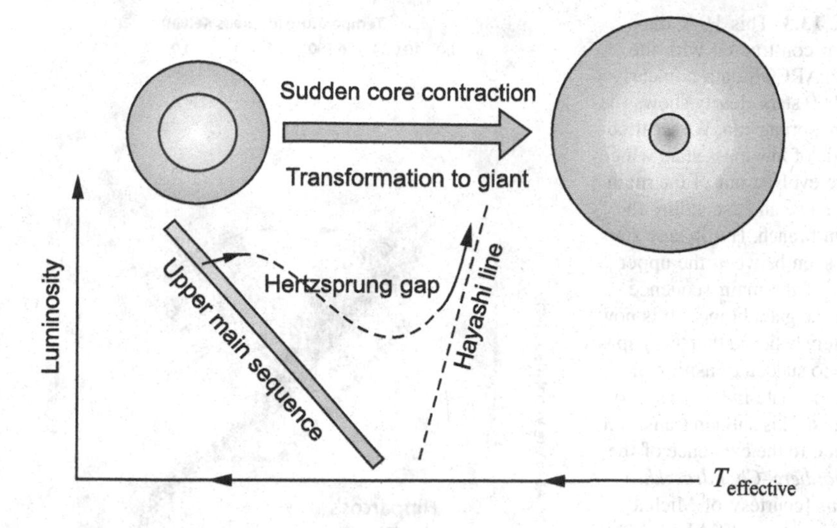

Fig. 13.2 This figure shows the trajectory of an intermediate mass star in the H–R diagram as the core reaches the *Schönberg–Chandrasekhar Limit*. Originally the star is close to the main sequence. With sudden contraction of the core, and the consequent expansion of the envelope, the star moves very rapidly towards the Hayashi line. From then on, it ascends the giant branch. Since the transition to a giant is very rapid—it happens on a thermal timescale, rather than a nuclear timescale—we are unlikely to catch many intermediate mass stars in the process of this transition. Indeed, very few stars are actually seen in the region of the H–R diagram marked as the *Hertzsprung gap*

in the shell, the location of the core in this diagram will move along this upper branch, maintaining equilibrium with the envelope. This will proceed continuously till the core mass reaches the *turning point* q_{SC}, which is the Schönberg–Chandrasekhar Limit. *When the mass of the core exceeds this critical value, the only equilibrium models are in the lower branch, and the core will have to contract discontinuously.* The sudden contraction of the core will be accompanied by an expansion of the star, and the star will move rapidly in the H–R diagram from the vicinity of the main sequence to the region of the Hayashi line. This is shown in Fig. 13.2.

This central conclusion, namely that the core will contract in an abrupt manner, and the star will expand to become a giant, is borne out by detailed numerical calculations done in recent years. The core contraction and the expansion of the star happen in a very short time $\sim$ a few *million* years. This is to be contrasted with *billions* of years for a star like the Sun. Since stars of intermediate mass evolve from the main sequence to the giant branch rather quickly, one would not expect to catch them during this transformation.

Indeed, there is a region in the H–R diagram known as the *Hertzsprung gap* in which there are very few stars. This may be seen in Fig. 13.3 (which is a reproduction of Fig. 12.11) which shows the H–R diagram plotted with the data obtained from the HIPPARCOS satellite. Whereas we clearly see the stars of roughly the mass of the

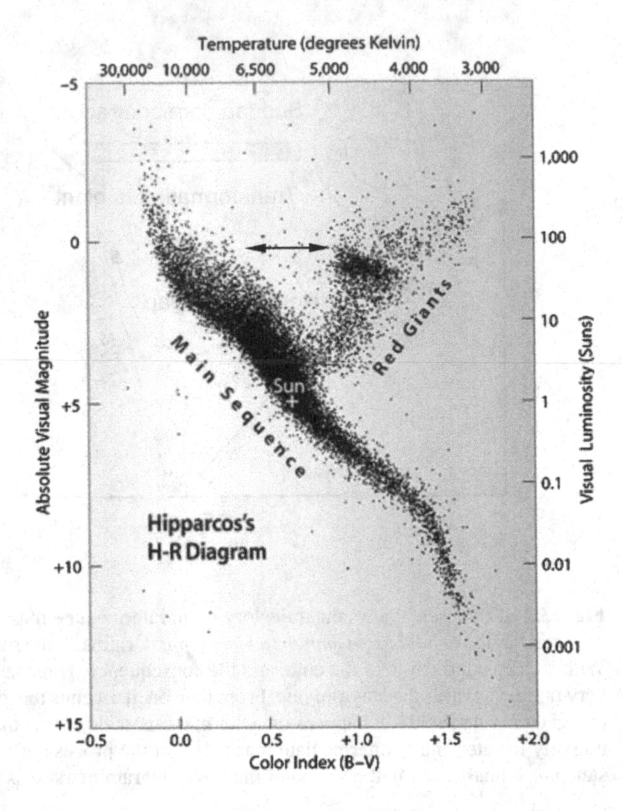

Fig. 13.3 This H–R diagram constructed with the HIPPARCOS data on nearly 23,000 stars clearly shows the Hertzsprung gap. We see thousands of low mass stars which have evolved out of the main sequence and ascending the giant branch. Hardly any stars are seen between the upper part of the main sequence and the giant branch. It is now widely believed that this gap is due to sudden transition of the intermediate mass stars into giants. This sudden transition is due to the existence of the *Schönberg–Chandrasekhar Limit* [courtesy of Michael Perryman and ESA]

Sun evolving from the main sequence and ascending the giant branch, we do not see the more massive stars evolving into giants.

This had been a puzzle for a long time. In contemporary literature, the sudden contraction of the core when the mass of the core reaches the Schönberg–Chandrasekhar Limit, and the consequent expansion of the star, is taken as the explanation of the *Hertzsprung gap.*

Central Helium Burning

The rapid contraction of the core will result in the heating of the core. When the temperature reaches 10^8 K, helium will start to fuse in the core. For a star of 5 $M_\odot$ this will happen at the age of approximately 60 million years. This is a relatively short time compared to 8 *billion* years for a 1 $M_\odot$ star; the Sun will burn hydrogen in the core for another 3 billion years! Further, since the helium core of the intermediate mass stars will not be degenerate, there will be no *helium flash*; the safety valve will be on and helium burning will be quiescent. When helium burning sets in, the star will be a red giant located close to the Hayashi line. One would, therefore, expect the star to be highly convective, and detailed calculations bear this out. The outer convection zone penetrates very deep into the star. *The larger the mass of the star, the deeper the convection zone penetrates*. This large-scale convection enables the nuclear species produced near the centre to be *dredged up to the surface;* convection provides a *conveyer belt*, so to speak, connecting the surface to the deep interior. Once these nuclei are present near the surface, they can be seen and studied spectroscopically.

You will remember that while on the main sequence (when hydrogen burns in the core) stars in the upper main sequence will have radiative envelopes, but convective cores. The core will continue to be convective when helium burns in the core. The high temperature sensitivity of helium burning causes a strong concentration of energy generation, leading to a steep temperature gradient in the core. This, in turn, results convective instability of the core. Now, since the star is near the Hayashi track, the outer layers will also be convective.

As we discussed in an earlier chapter, the triple alpha reaction produces carbon: $3\alpha \rightarrow {}^{12}C$. As the abundance of carbon increases, oxygen begins to form through the reaction ${}^{12}C + {}^{4}He \rightarrow {}^{16}O + \gamma$. As ${}^{4}He$ gets depleted, the formation of ${}^{16}O$ becomes more dominant than the formation of ${}^{12}C$. Calculations show that *when all the helium in the core is exhausted, there is roughly equal abundance of* ${}^{12}C$ *and* ${}^{16}O$. For a 5 $M_{\odot}$ star, the helium burning phase lasts about 10 million years, which is roughly 20 % of the main sequence phase.

The Carbon–Oxygen Core

When helium is exhausted in the central region, a dense core consisting of carbon and oxygen, roughly in equal proportion, is formed. Helium will continue to burn in a concentric shell surrounding the inert C–O core. Further out, there will be a shell in which hydrogen will burn for a while. The luminosity of the star will be due to these two shells. The helium burning shell will add more carbon and oxygen to the core. As the mass of the core increases, it will contract. The *mirror principle* will come into effect once again. The contraction of the core will result in the expansion of the star. The contraction of the core will also result in the increase in luminosity of the helium burning shell. As a result, the luminosity of the star will increase nearly tenfold.

As the core contracts, it will move progressively towards the right in the $\log T - \log \rho$ plane, and will soon become degenerate. This is shown in Fig. 13.4.

To Be or Not to Be!

The carbon–oxygen core becoming degenerate is not necessarily the end of the story for these intermediate-mass stars. We saw in the previous chapter that the cores of stars in the lower main sequence ($M < 2.5\ M_{\odot}$) become degenerate when hydrogen is exhausted in the core. Therefore, when Helium eventually ignites in the core, the

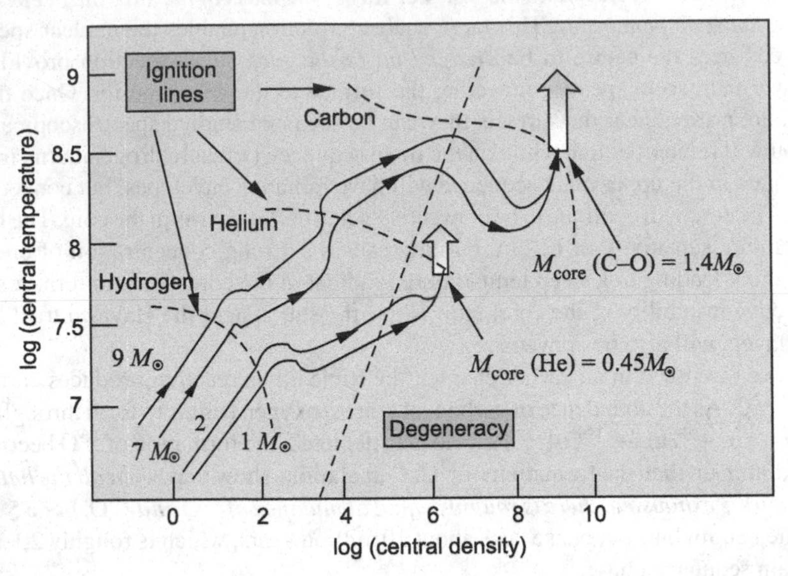

Fig. 13.4 Calculated evolutionary trajectories of the cores of low mass and intermediate-mass stars are schematically shown in this figure (This figure has been adapted from the textbook, *Stellar Structure and volution* by Kippenhahn and Weigert). In the case of the low-mass stars, the helium cores become degenerate. Helium ignites only when the mass of the degenerate core grows to 0.45 $M_{\odot}$. This results in a runaway energy generation for a while. These stars are saved because the core, once again, moves to the left of the dashed line, and becomes nondegenerate. In the case of intermediate-mass stars, helium cores remain nondegenerate. However, their carbon–oxygen cores do become degenerate. *Because of severe cooling due to neutrino emission, a degenerate carbon core will ignite only when the mass of the core reaches* 1.4 $M_{\odot}$. If the core does reach this mass, there will be an explosive carbon detonation, and the whole star will be blown apart!

core will be in a degenerate state. Consequently, there will be a *thermal runaway* (see Fig. 12.5). Fortunately, because of the rapid heating the core will cease to be degenerate and the safety valve will turn on before any damage is done (see Fig. 12.6 and the related discussion). *Calculations show that the degenerate helium core will ignite when its mass grows to* 0.45 $M_{\odot}$, *regardless of the mass of the star.* This is schematically shown in Fig. 13.4.

Let us return to our discussion of the *carbon–oxygen cores* of intermediate-mass stars. If helium did not continue to burn in a shell surrounding the core, it *will* be the end of the story for the star. The degenerate core will cool down at constant density and become a white dwarf. But the star need not have such a peaceful death! Since the helium burning shell will produce more and more carbon and oxygen, the mass of the core will increase. Because the core is now *degenerate, its radius will decrease as its mass increases* (recall the *inverse* relation between mass and radius of degenerate stars). More importantly, the gravitational energy released in the contraction can heat up the core—if there is no mechanism for heat loss.

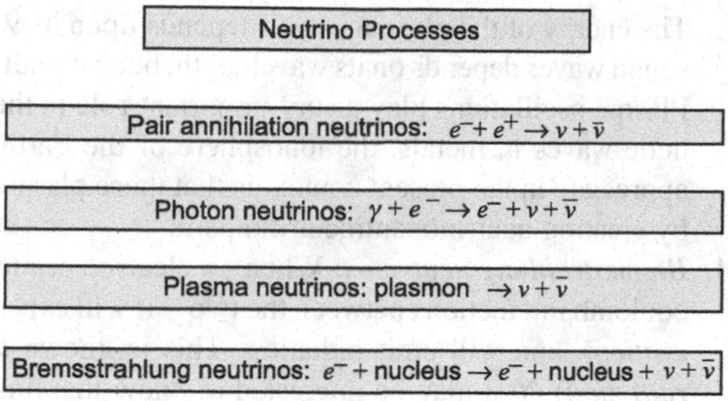

Fig. 13.5 At very high temperatures, neutrinos are produced in a variety of processes summarized here

Neutrino Processes

Pair annihilation neutrinos: $e^- + e^+ \rightarrow \nu + \bar{\nu}$

Photon neutrinos: $\gamma + e^- \rightarrow e^- + \nu + \bar{\nu}$

Plasma neutrinos: plasmon $\rightarrow \nu + \bar{\nu}$

Bremsstrahlung neutrinos: e^- + nucleus $\rightarrow e^-$ + nucleus + $\nu + \bar{\nu}$

There is an important new cooling mechanism at very high density and temperature. At temperatures less than about 10^8 K, and density less than about 10^7 g cm^{-3}, the main mechanism for cooling is the emission of photons. At higher temperatures and densities, emission of neutrinos is the main cooling mechanism. And it is very effective, since the neutrinos escape very easily. So far we have encountered the elusive neutrino only in the context of nuclear reactions. We first encountered it in the context of beta decay, and later in the context of various fusion reactions. But there are other channels for the copious production of neutrinos—channels unrelated to nuclear reactions. These are summarized in Fig.13.5.

1. *Pair annihilation neutrinos*: At temperature greater than $\sim 10^9$ K, photons in the tail of the black body spectrum can produce *electron–positron pairs*. The condition is that the energy of the photon must be greater than twice the rest mass energy of the electron: $h\nu \geq 2mc^2$ (remember that the electron and positron have the same mass). These electron–positron pairs will soon annihilate, with each such annihilation giving rise to two or three photons. There is, however, a very small probability that the annihilation will result in a *neutrino–antineutrino pair*. This will happen once in about 10^{19} annihilations. This is a very important process under the conditions we are considering.
2. *Photon neutrinos*: $\gamma + e^- \rightarrow e^- + \nu + \bar{\nu}$. You may recall the process known as Compton scattering. An x-ray or gamma ray photon scatters off an electron and changes its energy. In very rare cases, the scattered photon is replaced by a neutrino–antineutrino pair.
3. *Plasma neutrinos*: We have encountered the electron gas before. But we have not had the occasion to discuss an important property of such a gas. The electrons, in a background of positive charges (to make the system electrically neutral) can undergo *collective oscillations*. These collective oscillations are known as *plasmons*. The characteristic frequency of these quantum oscillations is given by

$$\omega_p^2 = \frac{4\pi n e^2}{m_e}.$$

The energy of the plasma waves depends upon its wavelength; just the energy of sound waves depends on its wavelength, but we shall not digress into those details. Plasma oscillations play a very important role in the propagation of electromagnetic waves in metals, the ionosphere of the Earth etc. The important thing to appreciate in the present context is that these plasma oscillations can lose energy by creating neutrino–antineutrino pairs.

4. *Bremsstrahlung neutrinos*: When an electron scatters off a nucleus—due to the coulomb interactions between the two—it will experience *acceleration* (or *deceleration*), and will emit radiation. This is known as *Bremsstrahlung* (or *brake radiation*). You may be interested to know that this is how x-rays are produced in the x-ray generators that one finds in hospitals. *At very high temperature and densities, the decelerating electron can create a neutrino-antineutrino pair.*
   All these processes are very temperature sensitive, with the *neutrino luminosity* increasing dramatically as the temperature increases.

Let us now get back to our discussion of the carbon oxygen core which is growing in mass, contracting and heating up adiabatically. If its temperature reaches about 5×10^8 K, carbon can fuse. Unfortunately, if it does happen, there will be a thermal runaway (see Fig. 12.5) and the star will blow itself up! Therefore, whether the star survives or *detonates* itself by igniting carbon depends upon how effective the cooling of the core due to neutrino emission is. Clearly, there are two competing effects:

1. The energy released in the fusion reaction (which heats up the core).
2. The energy lost due to neutrino emission (which cools the core). The efficiency of this mechanism increases with increasing temperature.

Detailed calculations show that till the core attains a mass very nearly equal to $1.4\,M_\odot$, neutrino cooling dominates. In other words, *the cooling of the carbon–oxygen core due to neutrino emission will prevent the core attaining the temperature needed for the ignition of carbon as long as the mass of the core is less than about* $1.4\,M_\odot$ (see Fig.13.6). By the way, this value of the critical mass of the core has nothing to do with the *Chandrasekhar limiting mass*, which is equal to $1.44\,M_\odot$! This is just a near coincidence—or, is it?

If the star is more massive than $1.4\,M_\odot$, there is no reason why the carbon–oxygen core cannot reach this critical mass. *One would therefore expect all stars more massive than* $1.4\,M_\odot$ *to blow up*. Till the late 1980s, this was thought to be the mechanism of the so called Type I Supernova in which no stellar remnant is left behind.

Then came the evidence from different quarters that stars with mass less than about $9\,M_\odot$ will save themselves from such a carbon detonation catastrophe. What changed the whole scenario was the discovery of white dwarfs in young clusters of stars, such as the Pleiades shown in Fig. 13.7.

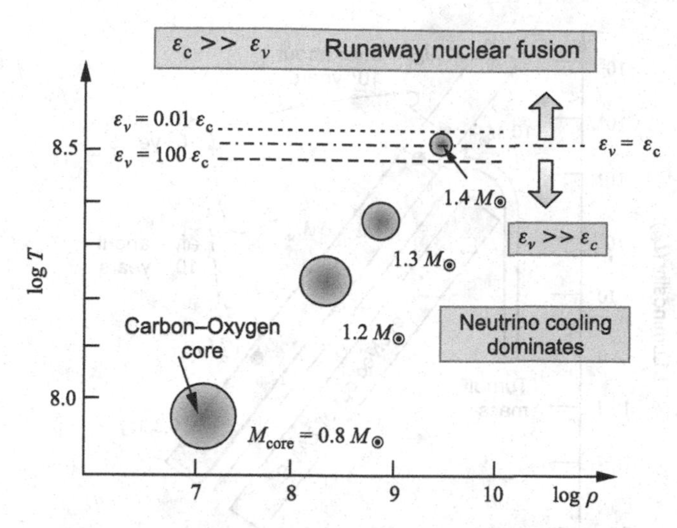

Fig. 13.6 As the mass of the degenerate carbon–oxygen core grows, it will contract (remember the inverse relation between mass and radius for degenerate configurations). The gravitational binding energy released will heat the core. So the core will move in a northeasterly direction in this plot. Even as the core heats up, cooling due to neutrino emission becomes more and more dominant. Indeed, if carbon was to ignite, the sudden heating of the core due to the energy released will result in such a dramatic increase in the cooling due to neutrinos, that the carbon burning will be quenched. In other words, cooling due to neutrinos dominates over the energy that would be released by fusion reactions. This remains so till the core grows to 1.4 $M_\odot$. Beyond this critical mass, energy released by carbon burning exceeds the energy loss due to the neutrinos. Unfortunately, carbon burning will be runaway reaction, and the star will be blown apart

Fig. 13.7 The **Pleiades**, or *Seven Sisters*, is an open star cluster containing middle-aged hot stars located in the constellation of *Taurus*. It is among the star clusters nearest to Earth and is the cluster most obvious to the naked eye in the night sky. Pleiades has several meanings in different cultures and traditions

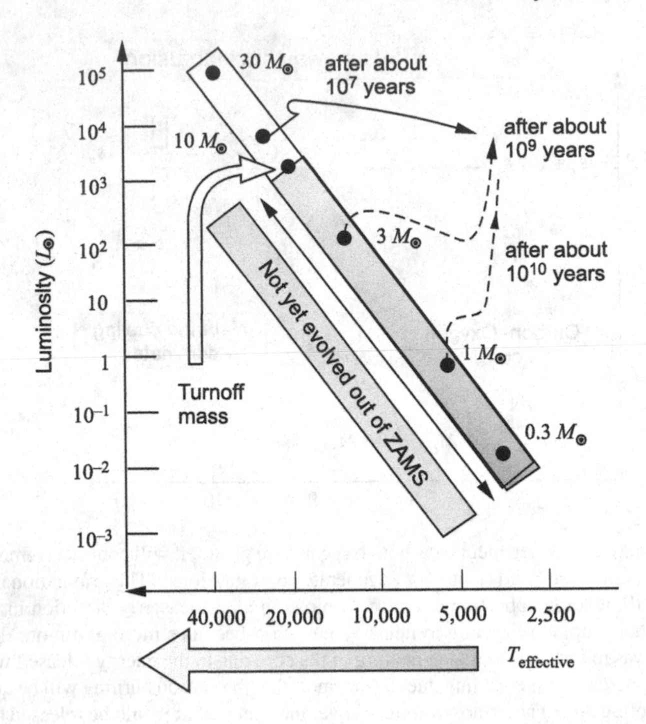

Fig. 13.8 The H–R diagram of a cluster of stars. Remember that all the stars in a cluster are born at the same time. Since massive stars evolve faster, the more massive ones would have evolved out of the main sequence and ended their lives. Only stars whose main sequence lifetime is greater or equal to the age of the cluster would still be on the main sequence. The maximum mass of the stars on the main sequence is known as the *turnoff mass*. In a young cluster like the Pleiades, the turnoff mass is as high as 6 or 7 $M_\odot$. And yet, there are white dwarfs in these clusters. This is only possible if stars more massive than the turnoff mass somehow manage to end their lives peacefully as white dwarfs!

The Pleiades cluster is at a distance of about 350 light years and contains roughly 1,000 stars. The cluster is dominated by hot blue and extremely luminous stars that have formed within the last 100 *million years*. It is thus a *young cluster of stars* presumably born in the same birth event. *And yet, there are many white dwarfs in the cluster.* It is fairly certain that these white dwarfs are original cluster members. If we accept the notion that white dwarfs are end states of stars less massive than the Chandrasekhar mass limit of 1.4 $M_\odot$, then we have a serious dilemma. We have seen that a low mass star like our Sun will take many *billions of years* to evolve from the main sequence. And yet, the cluster is only about 100 million years old! How could the white dwarfs have formed?

Figure 13.8 shows the Hertzsprung–Russell diagram for a young star cluster like the Pleiades. Notice that only stars in the upper part of the Zero-age Main Sequence (ZAMS), *with mass greater than the turnoff mass*, have evolved out of the main sequence, and ended their lives. For star with mass less than the *turnoff mass*, the

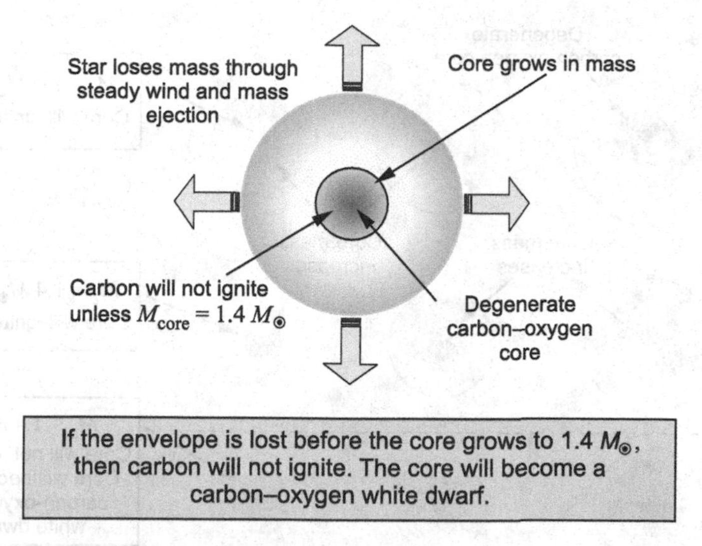

Fig. 13.9 Illustration of mass loss from the star

evolution time on the main sequence far exceeds the age of the cluster; they are, therefore, still on the main sequence. The turnoff mass will, of course, depend upon the age of the cluster. The older the cluster, the smaller will be the turnoff mass. *Globular clusters*, for example, are as old as the Galaxy itself (many billions of years old). In these clusters, one finds only very low mass stars. All other stars have had enough time to evolve and find ultimate peace as white dwarfs or neutron stars. In young clusters like the Pleiades and Hyades, the *turnoff mass* is around 6 or 7 $M_{\odot}$. The meaning of this is clear: Only stars more massive than, say, 7 $M_{\odot}$ have had time to evolve. *It follows therefore that the white dwarfs in these clusters are the end states of stars more massive than the turnoff mass*!

How could this be? A star of, say, 8 $M_{\odot}$ should have blown itself up. The carbon–oxygen core of such a star could have easily grown to a mass of 1.4 $M_{\odot}$, at which point carbon would have ignited, resulting in a detonation of the star. *The only way to avoid this is if the star lost a great deal of mass and thus prevented the core from growing to the critical mass*. This is shown schematically in Fig. 13.9. In recent times, there is observational evidence for mass loss from stars. This could be due to a variety of reasons:

1. Strong winds from the surface.
2. Mass loss due to rapid rotation of the star.
3. Periodic mass ejection during thermal pulsation of the star.
4. The sudden core contraction, resulting in a sudden expansion of the envelope (recall the Schönberg–Chandrasekhar limit), etc.

Figure 13.10 shows the evolution of stars of three different initial masses. Here, $M_1 > M_2 > M_3$. The degenerate carbon–oxygen core grows in mass, even as the star loses mass from the surface. In stars more massive than M_2 the core will grow

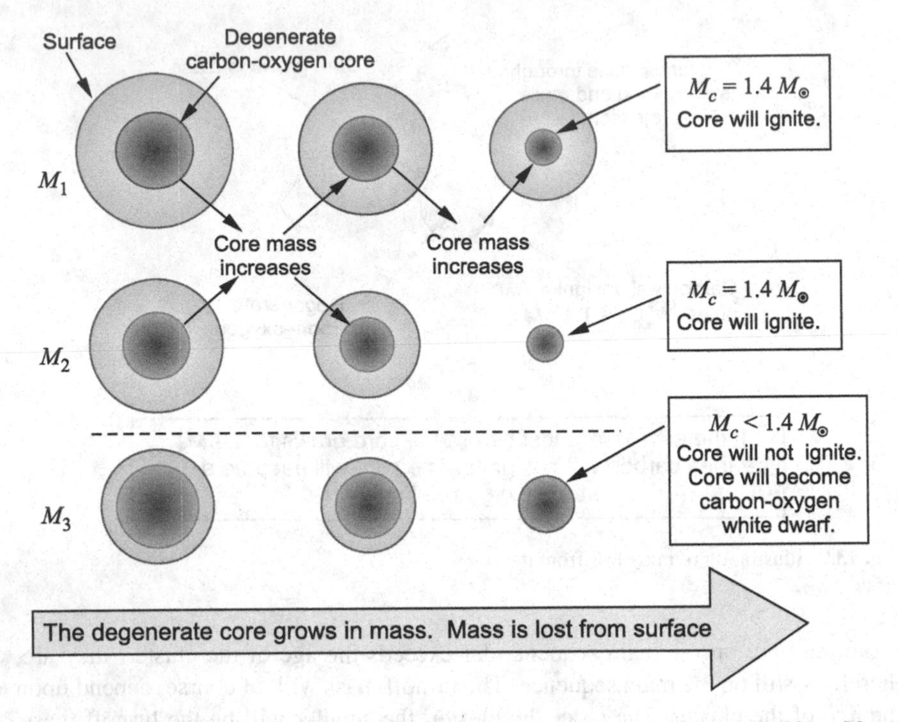

Fig. 13.10 A time sequence of the evolution of stars with three different initial masses; $M_1 > M_2 > M_3$. Even as the degenerate carbon-oxygen core increases in mass, the star is losing its envelope due to a variety of processes. Notice that as the mass of the core increases, its radius decreases. If the star manages to lose all its envelope before the core mass reaches the critical value of 1.4 $M_\odot$, the carbon will not ignite, and the *core will become a white dwarf*. Thus, *stars less massive than* M_3 *will die peacefully as carbon–oxygen white dwarfs*

to the critical mass before the envelope is completely lost. Carbon will ignite in the cores of these stars, resulting in an explosion of the star. In stars less massive than M_2, the envelope is lost before the core grows to the critical mass. Such core will cool down at constant density and end their lives as white dwarfs.

The existence of white dwarfs in young clusters like Pleiades, Hyades, etc. strongly suggests that star with mass less than about 9 $M_\odot$ will find ultimate peace as white dwarfs (see Fig. 13.11). While one might argue whether the upper mass for the formation of white dwarfs is 8 or 9 $M_\odot$, there is compelling observational evidence that the upper mass limit is in this range.

The main conclusion of this chapter is that intermediate-mass stars with masses from 2.5 to 9 $M_\odot$ will find their ultimate peace as *carbon–oxygen white dwarfs*. In the previous chapter, we concluded that stars in the lower part of the main sequence will also die as white dwarfs. Stars with mass less than about 0.5 $M_\odot$ will not be able to ignite helium formed during the main sequence phase; their cores will never get hot enough for this to happen. If they manage to get rid of the hydrogen envelope they will end up as *helium white dwarfs*. But the catch is that the evolution time for a

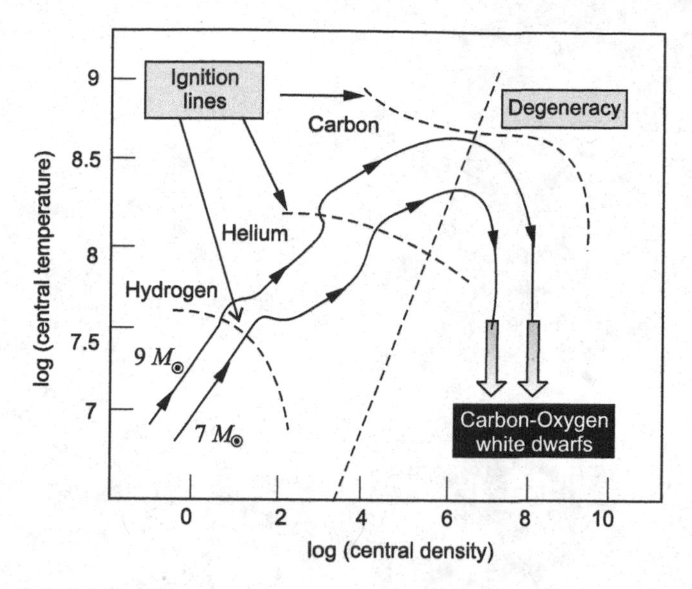

Fig. 13.11 Observations of white dwarfs in young star clusters like the Pleiades and Hyades strongly support the conclusion that stars with mass up to about 9 $M_\odot$ end their lives as carbon–oxygen white dwarfs. The stars manage to lose enough mass to prevent the core from growing to 1.4 $M_\odot$

star of such a low mass is more than the present age of the Universe! Therefore, the helium white dwarfs that we do occasionally find in the Galaxy must have formed via a different route. One possibility is that the star lost its entire hydrogen envelope to a close companion. But we shall not get into all those details.

The important thing is that all stars up to 9 $M_\odot$ will find ultimate peace. *They will have enough energy to cool!*

Chapter 14
Diamonds in the Sky

The Population of White Dwarfs

The final conclusion regarding the ultimate fate of stars less massive than about 9 $M_\odot$ is summarized in Fig. 14.1. It would be interesting to know what fraction of stars end up as white dwarfs. For this, we have to go to the *Initial Mass Function* (IMF) of stars in the galaxy. The IMF, usually denoted by $\Psi(M)dM$, is *the number of stars formed per year per cubic parsec within an interval of mass between M and* $M + dM$. A *parsec* is the distance to the star whose parallax angle is 1 *second of arc* (see Fig. 12.10); *one parsec is roughly equal to three light years.* In 1955, Edwin Salpeter found that

$$\Psi(M)dM = 2 \times 10^{-12} M^{-2.35} dM \text{ stars/year/cubic parsec.} \tag{14.1}$$

This famous *Salpeter Initial Mass Function* is sketched in Fig. 14.2. This function will enable us to determine what fraction of stars end up as white dwarfs, and what fraction end up as neutron stars or black holes. The ratio of the area under the curve from $0.5M_\odot$ to $9M_\odot$ to the area under curve from $9M_\odot$ to ∞ will gives us the ratio of white dwarfs to neutron stars plus black holes,

$$\frac{\text{number of proto white dwarfs}}{\text{number of proto NS/BH}} = \frac{\int_{0.5}^{9} \Psi(M)dM}{\int_{9}^{\infty} \Psi(M)dM}. \tag{14.2}$$

A simple exercise will tell you that 98 % of all stars are or will become white dwarfs. It is equally interesting to ask what fraction of *mass* is locked up as white dwarfs. To get the answer to this, all we have to do is to multiply the IMF by the mass M of the star and integrate. Try to convince yourself that roughly 94 % of all matter that formed stars is either already locked up as white dwarfs, or is in stars that will eventually become white dwarfs. Only 6 % of the mass is in the form of neutron stars, or is in stars that will end their lives as neutron stars or black holes.

G. Srinivasan, *Life and Death of the Stars*, Undergraduate Lecture Notes in Physics,
DOI: 10.1007/978-3-642-45384-7_14,

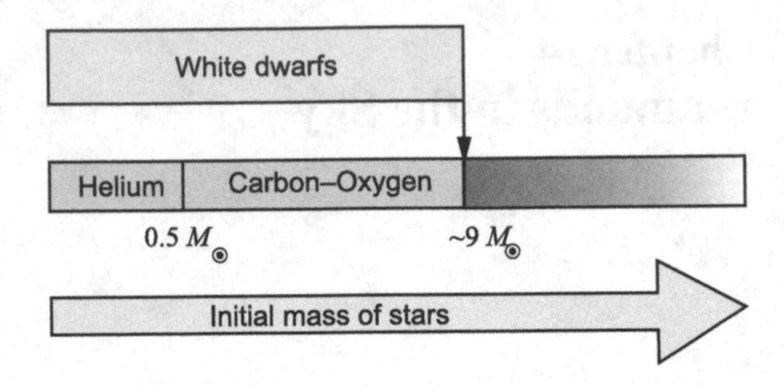

Fig. 14.1 This figure summarizes the final conclusion regarding the ultimate fate of stars less massive than about 9 $M_\odot$. While stars less massive than about 0.5 $M_\odot$ can, in principle, end up as helium white dwarfs, the majority will die as carbon–oxygen white dwarfs

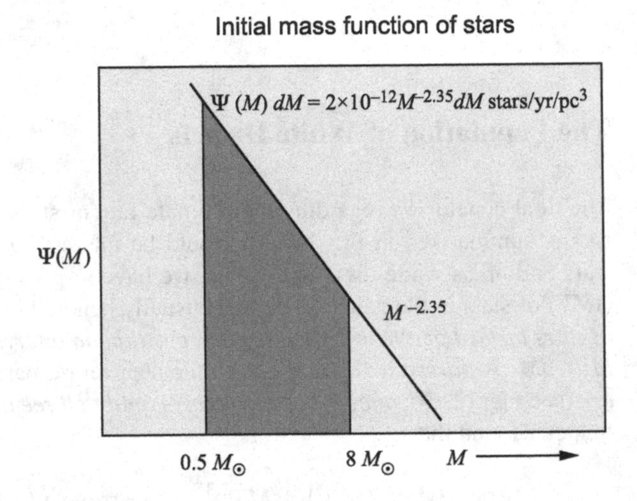

Fig. 14.2 Figure shows a sketch of the Salpeter Initial Mass Function. $\Psi(M)dM$, is the number of stars formed per year per cubic parsec within an interval of mass between M and $M+dM$. The *shaded area* will give us the fractions of all stars that have either ended their lives as white dwarfs, or will eventually end up as white dwarfs

This has an interesting implication. At the present age of our Galaxy, roughly half the matter is in the form of stars, and the other half is in the form of giant interstellar clouds of gas. Stars continue to form from these clouds of gas. But stars also *return* gas to the interstellar medium. However, it is only the more massive stars that explode as supernovae, and return most of their mass back to the Galaxy. Since the initial mass function of stars has a negative slope, in each generation of star formation the majority of newly born stars will be low-mass stars and therefore the majority of stars will end their lives as white dwarfs.

Eventually, there will be no interstellar gas left in the Galaxy! And the only stars that will be left behind will be very low mass stars which have still not evolved. Are there such systems? Yes, indeed. Our Galaxy contains a few hundred very old stellar systems known as *globular clusters*. Typically, they contain about a million stars, going around a common centre of mass. These globular clusters have hardly any gas. And their stellar population consists almost exclusively of very low mass stars. The more massive stars have all evolved and ended their lives, leaving behind white dwarfs, neutron stars and black holes. Another type of stellar systems with hardly any gas left are the so called *elliptical galaxies*. These are not *flat* like the spiral galaxies (Our own Milky Way Galaxy is an example of a spiral galaxy), but more like a *rugby ball* (an ellipsoid). These galaxies have very little gas left; and only very old, low-mass stars populate them.

Masses of White Dwarfs

Determination of the masses of white dwarfs relies on spectroscopy. Theoretically, a white dwarf should be a soup of atomic nuclei and degenerate electrons. But most white dwarfs have a very thin, and very pure, atmosphere. Observations tell us that this will be either hydrogen or helium. Emission lines from hydrogen atoms or helium ions can be detected in the spectrum of white dwarfs. These emanate from the thin outer layer. The mass of this layer is, however, extremely small, amounting to only about 10^{-4} to $10^{-3} M_{\odot}$. One of the techniques employed to determine the mass of a white dwarf is to determine the gravitational redshift of the wavelength of the spectral lines emanating from the surface (see Chap. 3, 'The strange companion of Sirius'). This, used in conjunction with Einstein's formula for the redshift, will give us the combination of mass and radius. One can now use the empirical mass–radius relation for white dwarf, and estimate the mass. Although such an estimate is prone to error, one can use it to derive the distribution of masses of white dwarfs using a very large sample. This has been done.

An extremely interesting fact to emerge in the 1990s is that the mass distribution of white dwarfs is quite narrow! *The mean mass of white dwarfs is about* $0.6M_{\odot}$. *The width of the mass distribution is only about* $0.14M_{\odot}$. This raises the question 'Why such a narrow mass distribution?' We shall not digress to discuss this interesting question. But it would be worthwhile to at least indicate the line of thinking that provides a plausible answer. You will recall from our discussion in the previous chapter (see Figs. 13.9 and 13.10) that the progenitor star is losing mass from its surface, even as the carbon–oxygen core is growing in mass. The rate at which the star loses mass is related to the luminosity it generates. The luminosity, in turn, is determined by the rate at which the mass of the core grows. So you may be able to detect a vague connection between the rate at which the core grows in mass, and the rate at which the star loses mass. Such a *conspiracy* can result in a *convergent situation*, leading to an almost unique mass for the white dwarfs, with a relatively small spread.

Magnetic White Dwarfs

Fairly strong magnetic fields have been detected in white dwarfs. There are two primary ways of detecting magnetic fields: (i) Strong magnetic fields produce measurable *circular polarization* in the spectrum, (ii) Zeeman Effect—the splitting of spectral lines due to the magnetic field (you may like to refer to our discussion of

Zeeman Effect in ***What Are the Stars?***). At present there are about 50 magnetic white dwarfs, with fields in excess of 10^4 Gauss (which is 10,000 times bigger than the Sun's average magnetic field!). Roughly 15 of them have fields less than 10^7 G, an equal number have fields between 10^7 and 10^8 G. Interestingly, roughly 15 white dwarfs have field between 10^8 and 10^9 G. The magnetic axis is usually not aligned with the rotation axis of the white dwarf.

What is the origin of the magnetic field? Why do some white dwarfs have huge fields, while others have none? There is reasonable consensus that the observed magnetic fields are *fossil fields*, that is, they are not generated at present, but inherited from their progenitors. Large-scale magnetic fields are usually created by current loops. Such current loops are driven by convective motions of the charged fluid. The point to bear in mind is that such motions are unlikely in a white dwarf or a neutron star. Since there is a dense and degenerate electron gas in a white dwarf, the star will have extremely high thermal conductivity (terrestrial metals have high thermal conductivity for the same reason). Such a high conductivity will ensure that the white dwarf is essentially *isothermal;* that is, there will be no appreciable temperature gradients (this is the case with metals, too). You will recall from our earlier discussions, *strong temperature gradients are essential for convection to set in. And convective motions are needed for any dynamo action.* This is the reason why the observed fields have to be fossil fields.

How does this work? Most stars have magnetic fields. Some stars, known as *Ap stars*, have strong magnetic fields of the order of 1000 G. Such fields are presumable generated in their cores due to dynamo action. As the core of the progenitor star contracts and becomes degenerate, *there will be an amplification of the field due to flux conservation.* This is a consequence of what is known as *flux freezing*. You may recall from your solid state physics course that high thermal conductivity implies high electrical conductivity. The ratio of the thermal conductivity to the electrical conductivity is a constant, which is proportional to the temperature. This is known as Wiedemann–Franz's Law. In a medium of high electrical conductivity (such as a metal or a plasma), the magnetic flux will be *frozen in*. To put it differently, it will cost a lot of energy to separate the field from the conducting medium. *If we try to move the conducting material, the field will move along with it. Conversely, if we try to move the field, then the magnetic field will drag the medium with it*. This fundamental principle was first elucidated by the Swedish Physicist Hannes Alfvén, a discovery that earned him a Nobel Prize for Physics in 1970. Imagine that there is a magnetic field of strength B at the centre of a conducting sphere of radius R. An immediate consequence of flux freezing is that when the sphere contracts from a radius R_1 to R_2, $B \times 4\pi R^2 =$ constant. In other words,

$$B_1 R_1^2 = B_2 R_2^2. \tag{14.3}$$

As may be seen from (14.3), during contraction, the magnetic field gets amplified by a factor which is equal to the square of the ratio of the radii. It is therefore plausible that the observed fields of white dwarfs are fossil fields, amplified during the contraction of the core of the progenitor. As we shall see in the next volume, neutron stars have magnetic fields in excess of 10^{12} Gauss. The radius of a white dwarf is of the order of 10,000 km, while that of a neutron star is only of the order of 10 km. Therefore, if the end result of stellar evolution is a neutron star, then one can anticipate a further amplification of the fossil field by a factor of 10^6. That would nicely give us a field of the order of 10^{12} G!

Cooling of White Dwarfs

A topic of contemporary interest is the rate of cooling of white dwarfs. If one has a good theory of the cooling rate then one can estimate the ages of white dwarfs. Of particular interest is the age determination of the coolest white dwarfs (presumably the oldest). It has been said that 'the history of star formation in our Galaxy is written in the coolest white dwarfs'. *In other words, since the coolest white dwarfs are the remnants of the oldest stars in the Galaxy, by studying their statistics, one can hope to recreate the history of star formation rate.*

If the white dwarfs did not have a thin atmosphere, then the theory of cooling would be straightforward. As we have mentioned earlier, degenerate matter has very high thermal and electrical conductivity. It is therefore safe to assume that the white dwarf will be isothermal, namely, the temperature will be the same everywhere in the star at any given moment. Since there is no energy generation in a white dwarf, the energy radiated by it is only the fossil heat. In the absence of any atmosphere, a white dwarf will obviously radiate as a black body. Do not be confused by the term *black body*. In the present context, any opaque body in which matter and radiation have come to a thermal equilibrium will radiate as a black body. The spectrum of radiation emitted by it will be known as *black body radiation*. Further the total energy radiated per unit time (or luminosity) will be $L = (4\pi R^2)\sigma T^4$, where T is the temperature and R is the radius of the star. This will result in a decrease in the stored heat energy and, therefore, the temperature. This, in turn, will result in a decrease in the luminosity. The moral of the story is that *as the white dwarf cools, the rate of cooling will decrease*. This can be formally expressed by the following equation:

$$L \propto (\text{specific heat}) \times M \times \frac{\partial T}{\partial t}. \tag{14.4}$$

You will remember that the *heat capacity* of a body is the *specific heat* multiplied by the mass; *specific heat* is defined *per unit mass*. For a body like a white dwarf, there are two contributions to the specific heat: the degenerate electron gas and the ideal gas of ions. We shall not go into the details here, but it turns out that the specific heat of the degenerate electron gas is much smaller than that of the ions. Another way of stating this is the following: the heat energy is mainly in the form of the motion of the ions. Although the degenerate electrons are moving like mad, that motion is zero-point motion, and nothing to do with heat. You will recall that because the ions are much more massive than the electrons, they can still be regarded as an ideal gas.

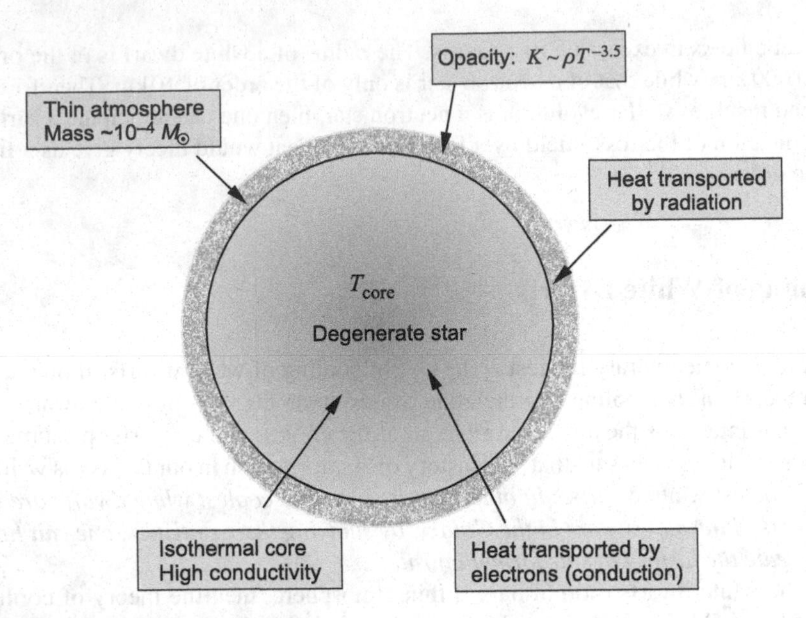

Fig. 14.3 A cooling white dwarf is like a hot metal with an insulating blanket around it. Whereas heat is transported extremely efficiently in the bulk of the star by the degenerate electrons, in the tenuous atmosphere, heat is transported by radiation (very much like in the envelope of the Sun). Like in a star, radiative heat transport is diffusive, and limited by the opacity of the matter. The lower the temperature, greater is the opacity

The specific heat of an ideal gas (at constant volume) is independent of temperature, and given by a very simple expression: $c_V = \frac{3}{2}Nk_B$, where N is the number of ions. This is the well-known Dulong and Petit's law.

But our problem is a little more complicated (see Fig. 14.3). White dwarfs do have an atmosphere. Although the mass of the atmosphere is only of the order of 10^{-4} $M_{\odot}$, it is like an *insulating blanket* around the white dwarf. Heat transport in the atmosphere is by radiation itself. And this is a diffusive process. The mean free path of the photons is governed by the opacity of the atmosphere. The various absorption and scattering processes that photons encounter in a star will, again, be operative here. We shall not go into the details here, but merely say that the *opacity* or *obstructive power* of the atmosphere will increase as the white dwarf cools.

The internal temperature of an infant white dwarf could be as high as 10^7 K. At these temperatures, cooling by neutrino emission is more effective than cooling by photons. This will be so for the first 10^7 years or more. After that photons take over from the neutrinos. During this later phase, the effective surface temperature of the white dwarf may be much less than its internal temperature. When the luminosity of the white dwarf has declined to about $10^{-4}L_{\odot}$ (by this time the effective surface temperature would have dropped to less than 10,000 K) the ions are expected to solidify and *crystallize*. This, too, has implications for the cooling of white dwarfs. *When ions crystallize, the latent heat released will temporarily heat up the white dwarf!*

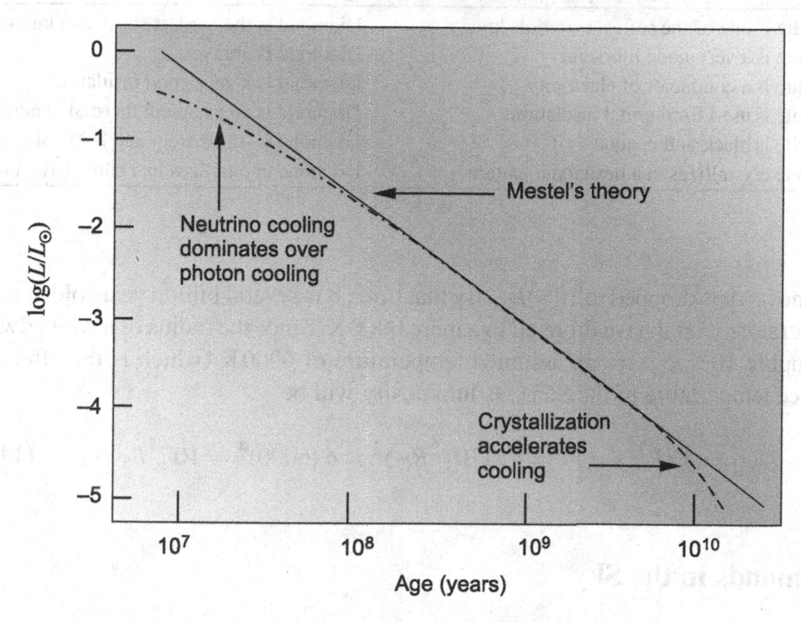

Fig. 14.4 The cooling curve of white dwarfs. Mestel's theory (the *solid line*) ignores the role of neutrinos in the cooling process. It also ignores the effects of crystallization of the interior. It is important to note that neutrinos contribute to the cooling in a significant manner for the first ten million years or more

After the ions solidify, the specific heat is due to the vibrations of the lattice. As the solid cools, the vibrations of the ions about their equilibrium positions become less vigorous, and *the specific heat drops rapidly with decreasing temperature.* Why the specific heat of a solid decreases with temperature was a great puzzle at the beginning of the twentieth century. This puzzle was solved in 1907 by Albert Einstein by invoking the radical idea that the atoms in a solid are *quantum oscillators*. You will remember that in 1905 Einstein had introduced the notion that the energy of electromagnetic radiation is quantized. These two papers by Einstein laid the foundation for the quantum theory of matter and radiation. While the *photoelectric effect* provided the evidence for the quantum nature of radiation, it is the discovery of the *Raman Effect* (in 1928) that provided the evidence for the quantum nature of matter (C. V. Raman was awarded the Nobel Prize for Physics in 1930).

To get back to our white dwarf, the crystallization of its interior, and the consequent decrease in the specific heat, will result in a dramatic increase in the cooling rate. Lower specific heat means lower *heat capacity*, that is, smaller capacity to hold in the heat. In Fig. 14.4 we have schematically shown the cooling curve for a $0.6M_{.}$ carbon–oxygen white dwarf. The solid line is the standard theory that was originally advanced by Mestel in 1952. It ignores the role of the neutrinos, as well as the effects of crystallization. Notice that a white dwarf crystallizes only after its

Table 14.1 Properties of graphite and diamond

Graphite is one of the softest materials known	Diamond is the hardest substance known
Graphite is a very good lubricant	Diamond is abrasive
Graphite is a conductor of electricity	Diamond is a very good insulator
Graphite is used for thermal insulation	Diamond is an excellent thermal conductor
Graphite is black and opaque	Diamond is transparent and brilliant
Graphite crystallizes in a hexagonal pattern	Diamond crystallizes in a cubic structure

luminosity has dropped to $10^{-4}L_{\odot}$. By that time, it is several billion years old, and its temperature would have dropped to a mere 6000 K. Since the radius of a white dwarf is roughly $10^{-2}R_{\odot}$, for an assumed temperature of 6000 K (which is the effective surface temperature of the Sun), its luminosity will be

$$L_{\mathrm{WD}} = 4\pi R^2 \times \sigma T^4 = 4\pi(10^{-2}R_{\odot})^2 \times \sigma(6000)^4 = 10^{-4}L_{\odot}. \tag{14.5}$$

Diamonds in the Sky

The suggestion that very old white dwarfs will crystallize was made way back in the 1960s. A further interesting suggestion was made in the 1980s by a series of astronomers. They argued that in very old carbon–oxygen white dwarfs (>5 × 10^9 years), carbon and oxygen will *phase separate* before crystallization. In this scenario, oxygen will settle towards the centre of the star *like snow flakes falling to the ground*. As a consequence, when the interior solidifies, it will have a solid oxygen core, surrounded by an envelope of solid carbon.

Carbon occurs in many different forms. The most common and familiar among them are graphite and diamond. Interestingly, these two *phases* of carbon are as different as they can be! These differences are summarized in Table 14.1 and Fig. 14.5.

If graphite and diamond both consist of carbon atoms, can we convert graphite into diamond? Indeed, we can! All one has to do is to subject graphite to enormous pressure. To understand this, let us look at what is known as the *phase diagram*. Fig. 14.6 shows the theoretical phase diagram of carbon. Such a diagram tells you the range of temperature and pressure in which a particular phase is the stable equilibrium phase. The pressure is plotted along the vertical axis, in units of *giga pascal* (named after the French scientist Pascal). Atmospheric pressure, also called a *bar*, is equal to 10^5 Pascal. One giga pascal is 10^9 Pa, or ten thousand atmospheres.

You can see that if the temperature is less than about 4000 K, graphite can be transformed to diamond by applying sufficient pressure. This is how *artificial diamonds* are produced! These are seldom of gem quality. How does Nature do it? As we go down beneath the surface of the earth, the pressure increases. In regions where the temperature and the pressure are both in the correct range, carbon derived from buried organic matter got converted to diamond millions of years ago. Here is an

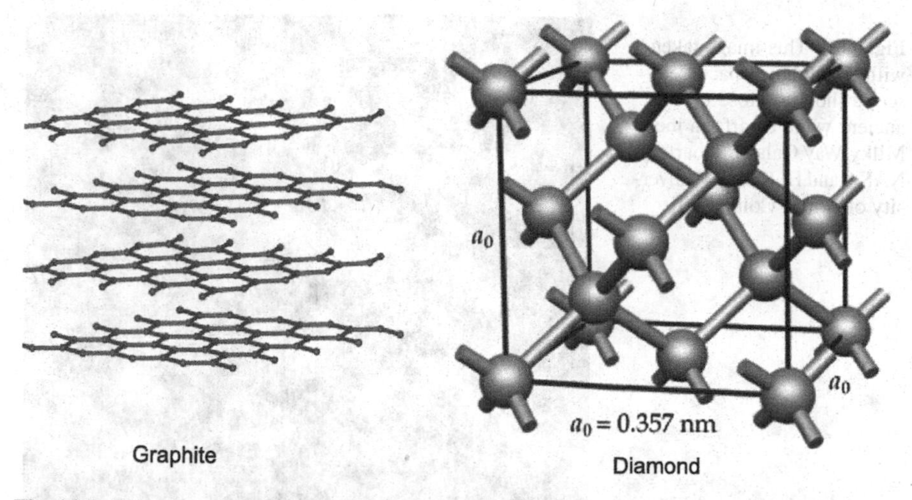

Fig. 14.5 The crystal structure of graphite and diamond

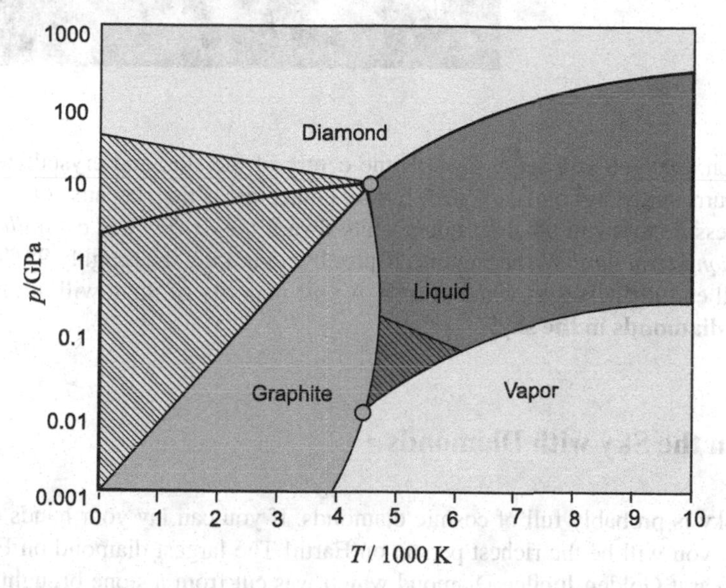

Fig. 14.6 A theoretical phase diagram of carbon. The *vertical axis* is the pressure and the *horizontal axis* is the temperature. Such a phase diagram shows the regions in which carbon exists as *vapour*, *liquid*, *graphite* and *diamond*

interesting question for you! When we mine the diamonds, and bring them to the surface, they are no longer subjected to immense pressure. *Why does the diamond not revert back to the state of graphite?* If you want to know more about this, I refer you to the fascinating monograph by G. Venkataraman, '*The Many Phases of Matter*'.

Let us get back to our ageing carbon–oxygen white dwarf. We were talking about the conjecture that when these white dwarfs crystallize, there will be a phase

Fig. 14.7 This image taken with the Hubble Space Telescope shows a close up of ancient white dwarfs in the Milky Way Galaxy [Courtesy NASA and H. Richer (University of British Columbia)]

separation. Oxygen will settle towards the centre of the star, and crystallize. This will be surrounded by solid carbon. A further conjecture is that, because of the enormous pressure that will obtain inside a white dwarf, *the carbon will crystallize into the diamond structure!* We began this chapter by remarking the roughly 98 % of the stars will end their lives as white dwarfs. A vast majority of them will, ultimately, become **diamonds in the sky!**

Lucy in the Sky with Diamonds

So the sky is probably full of cosmic diamonds. If you can lay your hands on one of these, you will be the richest person on Earth! The largest diamond on Earth is the 546-carat Golden Jubilee Diamond which was cut from a stone brought out of the Premier mine in South Africa. The cosmic diamond we are talking about is **10 billion trillion trillion carats!** (Fig. 14.7).

Astronomers may have finally found one of these! The white dwarf in question (discovered in 2004) is roughly 50 light years from the earth in the constellation Centaurus. Its official name is BPM-37093, rather unromantic. But astronomers have decided to call it **Lucy**, after the famous song by the iconic Beatles *Lucy in the Sky with Diamonds*. How do we know this white dwarf has crystallized?

Asteroseismology

Many white dwarfs pulsate; the intensity of light we receive from them shows periodic variations. These could be *radial vibrations* or more complicated vibrations (known as nonradial oscillations). By determining the frequency of these modes, one can infer the conditions that prevail inside the star. *This is just like trying to understand the properties of the material of a bell, or a gong, by studying its vibrational modes.* This is precisely how we infer the properties of the interior of the Earth. That is the discipline of *seismology*. If you have had the chance to read the previous volume in this series, ***What Are the Stars?*** you will remember that *helioseismology* has enabled astronomers to infer the conditions that prevail inside the Sun with incredible precision. In a similar manner, *asteroseismology* is the study of the pulsations of white dwarfs. Since white dwarfs are much farther away than the Sun, observations are more difficult. Nevertheless, astronomers have been able to deduce many properties of pulsating white dwarfs, such as its rotation rate. One of the questions that can be resolved by asteroseismology is whether the interior is fluid or a solid. Astronomers have come to the conclusion that *Lucy* has crystallized. There is some argument as to whether 90 % of the interior, or 75 %, has solidified. But there is consensus that a major fraction of the interior has solidified. If this conclusion is correct then *our understanding of the phase diagram of carbon enables us to say with some confidence that we have found one of the cosmic diamonds!*

Asteroseismology

Chapter 15
Exploding Stars

The Fate of Massive Stars

It now remains for us to discuss the evolution of the more massive stars, stars more massive than about $10M_\odot$. The evolution of these stars through the helium burning phase is more or less the same as that of the intermediate-mass stars which we discussed in Chap. 14, 'Life history of intermediate-mass stars'. The essential difference is that *in the massive stars the carbon-oxygen core never becomes degenerate during the contraction*. Figure 15.1 shows the computed evolutionary track for a $15M_\odot$ star.

Since the carbon core will be nondegenerate, when carbon does ignite it will do so in a quiescent manner. With the danger of the core becoming degenerate at last out of the way, the subsequent phases will also proceed in an uneventful manner. The end products of these phases will be *neon*, *oxygen*, and finally *silicon*. In the final nuclear cycle, *silicon* will fuse to form ^{56}Fe. As we discussed earlier, iron is the final stage for spontaneous fusion. The ^{56}Fe nucleus is the most strongly bound of all nuclei, and further fusion will cost energy, rather than release energy. After millions of years, the star has finally reached the end of the road.

Low mass stars like the Sun had to contend with a helium bomb. They had to defuse the bomb before the star blew up. Intermediate stars had to be very clever to avoid the ignition of carbon in their degenerate cores; neutrinos came to their rescue. A massive star does not have to be ingenious and shed much of its mass to prevent the ignition of carbon in the core. The nuclear history of massive stars is quite uneventful. We had anticipated this in our historical perspective. Let us recall Chandrasekhar's remarkable statement of 1932:

> For all stars of mass greater than M_{critical}, the perfect gas equation of state does not break down, however high the density may become, and the matter does not become degenerate.

Let us recall also the basis for this assertion. Chandrasekhar had shown that if radiation pressure exceeds 9.2 % of the total pressure, the ideas gas law will not break down. This is illustrated in Fig. 15.2, which has been adapted from Chandrasekhar's original paper of 1932. What one is attempting to do is to compare the degeneracy pressure of the electrons with the pressure calculated by assuming that the electrons

G. Srinivasan, *Life and Death of the Stars*, Undergraduate Lecture Notes in Physics,
DOI: 10.1007/978-3-642-45384-7_15,

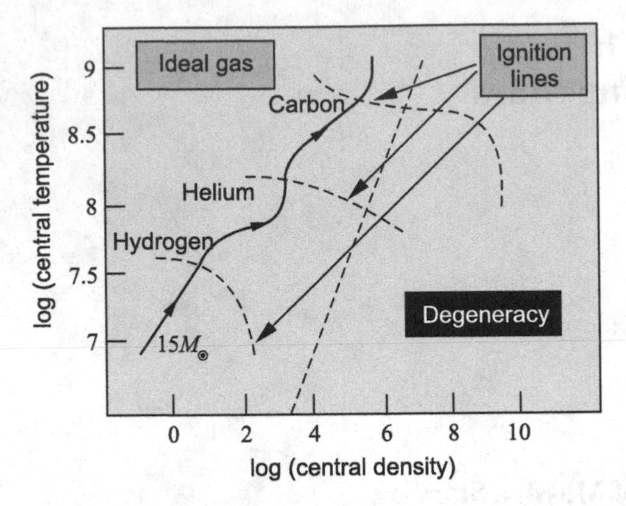

Fig. 15.1 This figure shows schematically the result of a modern evolutionary calculation for a $15M_\odot$ star. The *solid line* is the evolutionary track of the core of the star. The core remains nondegenerate through the successive phases of nuclear burning. Consequently, the fusion reactions proceed in a controlled manner, and the end product is a degenerate iron core. The figure has been adapted from the textbook, *Stellar Structure and Evolution*, by Kippenhahn and Weigert

obey the ideal gas law of Charles and Boyle, *with radiation pressure contributing in different measures.* The degeneracy pressure of the electron gas is given by $P = K_1\rho^{5/3}$ (nonrelativistic) and $P = K_2\rho^{4/3}$ (relativistic). These are the two thick lines. The lines labelled (1) to (4) represent the ideal gas equation of state of the electrons, with radiation pressure contributing to the total pressure in different measures. At first sight, you might be surprised that one can even plot the ideal gas equation of state in a (P, ρ)-plot since *the gas pressure is a function of both density and temperature.*

But this difficulty can be overcome by using the trick that Eddington introduced. One introduces a dimensionless fraction β defined as follows:

$$\boxed{\begin{aligned} P_{\text{tot}} &= \frac{1}{\beta}P_{\text{gas}} = \frac{1}{1-\beta}P_{\text{rad}} \\ &= \frac{1}{\beta}\left(\frac{\rho kT}{\mu_e m_p}\right) = \frac{1}{1-\beta}\left(\frac{1}{3}aT^4\right) \end{aligned}} \tag{15.1}$$

Equating the two sides of the second equation above and simplifying, one gets

$$\boxed{P_{\text{gas}} = C(\beta)\rho^{\frac{4}{3}} = \left[\left(\frac{k}{\mu_e m_p}\right)^4 \frac{3}{a}\frac{(1-\beta)}{\beta}\right]^{\frac{1}{3}} \rho^{\frac{4}{3}}.} \tag{15.2}$$

The above equation expresses the gas pressure in terms of the density and β. For intermediate steps leading to Eq. (15.2) refer to Chap. 7, in paticular, Eqs. (7.31) to

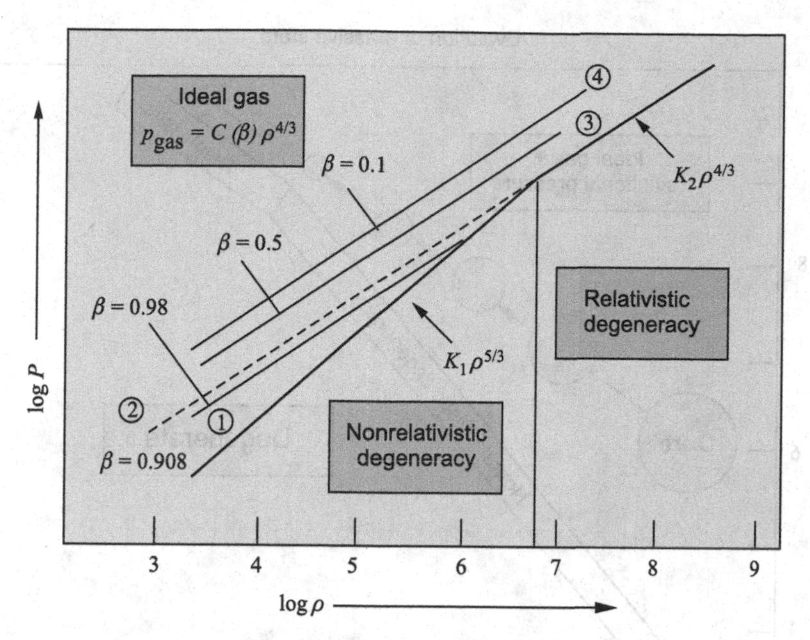

Fig. 15.2 The equation of state for an ideal gas (with radiation) is plotted as a function of the mass density. Although the pressure is a function of density *and* temperature, one can replace the temperature by the dimensionless fraction β, which is the ratio of gas pressure to the TOTAL pressure. The various lines of slope 4/3, and labelled ① to ④, represent the pressure of the ideal gas for various values of β. The pressure of a degenerate electron gas is shown by the two *thick lines*. As may be seen, *the dashed line which is the ideal gas with* $\beta = 0.908$, *that is, gas pressure is 90.8 % and radiation pressure is 9.2 % of the total pressure, respectively, is the borderline case. When radiation pressure exceeds 9.2 % of the total pressure* (or $\beta < 0.908$), *the ideal gas does not become degenerate no matter how high the density becomes*. This remarkable theorem was proved by Chandrasekhar in 1932. Indeed, this figure has been adapted from the famous paper published in *Zeitschrift für Astrophysik* in 1932

(7.34). The meaning of β is clear. β *is the ratio of gas pressure to the total pressure*. It follows that $(1 - \beta)$ *is the ratio of radiation pressure to the total pressure*. The lines ① to ④ have been labelled with the value of β. *Decreasing* β *signifies increasing proportion of radiation pressure*. Recall also Eddington's dictum: *radiation pressure is more important as we go to more massive stars*. Therefore, *the lines with decreasing* β *(or, equivalently, increasing radiation pressure) represent more and more massive stars*.

Let us now look at Fig. 15.2 closely. The line ① represents a star in which gas pressure contributes 98 % to the total pressure and radiation pressure contributes only 2 %. This line will intersect the nonrelativistic degeneracy pressure. Therefore, degeneracy will set in in such stars. Recall that *the condition for degeneracy is that the numerical value of the pressure calculated using Fermi–Dirac statistics is greater than the pressure calculated using Boyle's law* (at the same density and temperature). The line ② corresponds to *the critical value of radiation pressure equal to **9.2 % of***

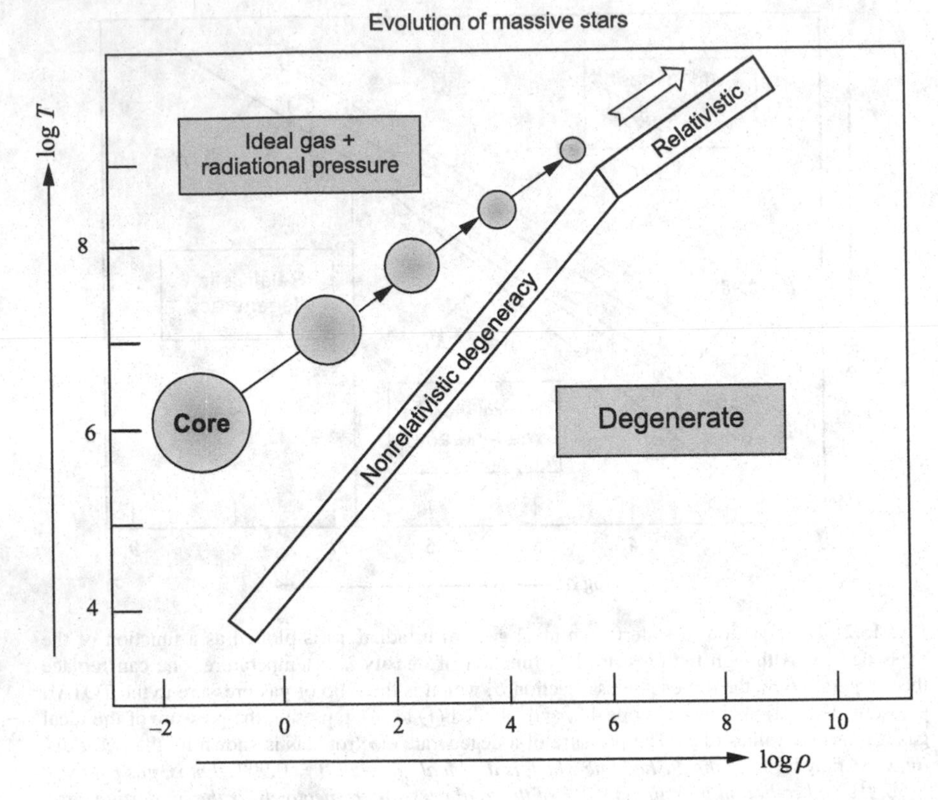

Fig. 15.3 The cores of stars more massive than about $10M_{\odot}$ never become degenerate, however high the density may become. In such stars, the nuclear fusion reactions proceed in a quiescent manner till an inert iron core is formed

the total pressure. This line not only does not intersect the nonrelativistic degeneracy line, it also misses the relativistic degeneracy line (both have a slope of $\frac{4}{3}$). This is, of course, true of even more massive stars in which $(1-\beta)$ is greater than 0.092, that is, radiation pressure exceeds 9.2 % of the total pressure. *In these stars, degeneracy will never set in, however high the density may become.* The evolution of the core of a massive star is shown in Fig. 15.3. A comparison with Fig. 15.1 shows that this conclusion dating back to 1932, based only on very general considerations, is fully borne out by modern calculations.

The Final Day!

The carbon burning phase lasts for a relatively short time in comparison to the helium burning phase, let alone the hydrogen burning phase. For a star of $15M_{\odot}$, the carbon burning phase lasts only about 6,000 years, while for a $25M_{\odot}$ star it lasts only for a

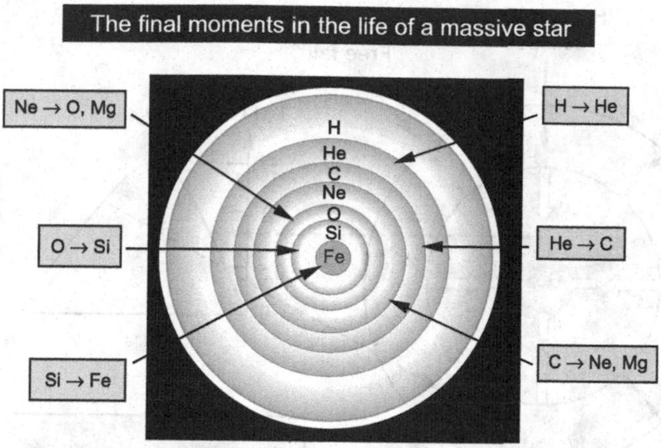

The final moments in the life of a massive star

Fig. 15.4 After an uneventful life of many millions of years, the fate of a massive star will be decided in an incredibly short time of the order of a day. On the final day of its life, Silicon will fuse in the innermost region to produce an iron core. Because of extremely efficient cooling due to neutrino emission, radiation pressure will, at last, become less than 9.2 % of the total pressure, and the core will become degenerate. And when the mass of this degenerate iron core grows to the Chandrasekhar limiting mass, it will collapse

hundred years! As we have already seen, during this phase there is strong neutrino emission. The *neutrino luminosity* of the star can equal, or even exceed, the photon luminosity of the star! Since the neutrinos escape, this results in the cooling of the core. But despite this the radiation pressure remains greater than 9.2 % of the total pressure, and consequently the core remains nondegenerate (Fig. 15.4).

The duration of the subsequent phases of the nuclear cycle are even shorter. The final act of the nuclear drama—the silicon burning phase—is extremely short. *For a 25 solar mass star, the silicon burning phase lasts only about a day!* The neutrino luminosity is incredibly large during this phase, roughly a million times the photon luminosity. The result of silicon burning is the formation of an iron core. Since the iron core is inert, it contracts to a density of about 10^{10} g cm^{-3}, with the central temperature of the order of 10^{10} K. *Cooling due to the neutrinos finally reduces the radiation pressure below the critical value of 9.2 %, and the iron core becomes degenerate at last!* It will now be supported by the degeneracy pressure of the electrons. So the star has an iron white dwarf at its centre. And the mass of this white dwarf will grow rapidly—as more silicon is converted to iron—till it reaches the Chandrasekhar limiting mass for white dwarfs. The electrons will be relativistically degenerate.

The Collapse of the Core

At this stage, the core becomes dynamically unstable and collapses. Historically, there have been many scenarios for why this collapse occurs. But the most persuasive one, at the time of writing this monograph, is the following. The electrons are relativistic

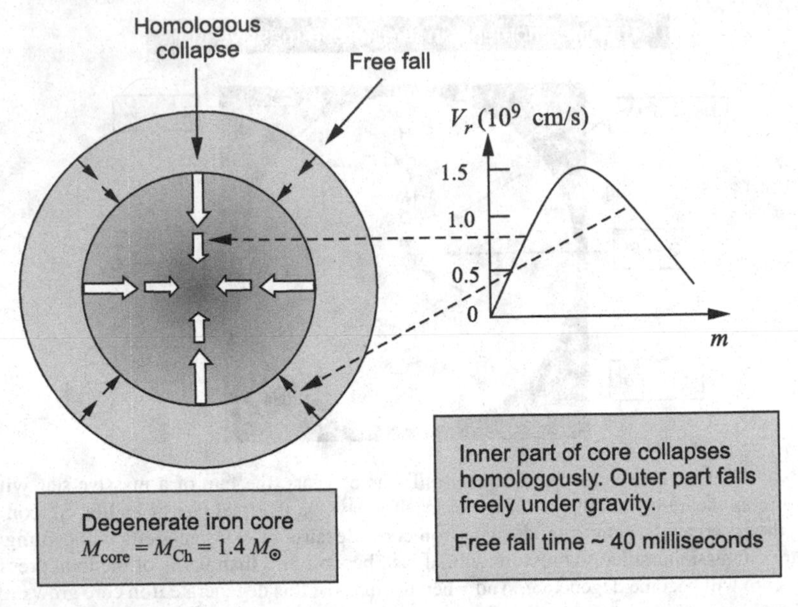

Fig. 15.5 This figure explains our present understanding of how the core collapses. The inner part of the core will collapse homologously (that is, preserving the radial density profile, even as it collapses), while the outer part of the core falls *freely*

and have a Fermi energy $E_F \approx 8$ MeV. This is the ideal condition for *inverse beta decay* (also known as *electron capture*) to set in. You will recall from our earlier discussion that in inverse beta decay an electron combines with a proton to form a neutron: $e^- + p \rightarrow n + \nu$. This *neutronization* of nuclei results in a rapid decrease in the number of electrons. Since the degeneracy pressure of the relativistic electrons is proportional to the density of electrons ($P_e \propto n^{\frac{4}{3}}$), there is a sudden diminishing of pressure. This will accelerate the collapse of the core. Clearly, this is a runaway process. There has been a great deal of effort to work out the details of this collapse, but it is a formidable problem. Let us try to understand this collapse at a basic level (see Fig. 15.5).

Calculations show that the *inner part of the core* collapses *homologously*. What this means is that during the collapse the radial density profile is maintained. If you think about it a little bit, you can convince yourself that in order for the radial density profile to remain the same, the outer shells have to fall in with greater speed than the inner shells. In contrast, the *outer regions* of the iron core are falling in with a radial velocity which decreases with increasing distance from the centre. This is just like a stone *falling freely* towards the Earth. The boundary between the inner and outer regions of the core is not rigidly fixed. As the collapse proceeds, there is less and less mass in the homologously collapsing region and more and more mass in the freely falling region.

The collapse is extremely short lived. The *free fall time* gives a good estimate for the duration of the collapse. The characteristic timescale for free fall is given by

$$\tau_{\rm ff} \approx \frac{1}{\sqrt{G\rho}}. \tag{15.3}$$

For an initial density of 10^{10} g cm^{-3}, one obtains a timescale of 40 milliseconds. For a density of 10^{14} g cm^{-3}, the free fall timescale is roughly 0.5 millisecond. Let us digress for a moment before proceeding. Consider the following problems: (i) A particle orbiting the Earth very close to the surface, (ii) A particle executing simple harmonic oscillations in a tunnel dug through the Earth, (iii) The vibration of the Earth. In all these cases, the characteristic timescale is $\sim(G\rho)^{-1/2}$. Think about why this is so!

Before outlining what happens during and after the collapse, let me tell you the end of the story first. The result of the collapse is the formation of a *neutron star*. The gravitational binding energy released in the process blows up the rest of the star, and this results in a *supernova*. Although this sounds plausible, the details of how the gravitational energy is actually utilized to produce an explosion are still somewhat uncertain. This is despite very many clever physicists working on this problem for several decades. Let us try to get a flavour of the complexity involved.

Let us first estimate the gravitational energy released in the collapse of the core. As already mentioned, the iron core has a mass roughly equal to the Chandrasekhar mass. Its density is of the order of 10^{10} g cm^{-3} and its radius is about a thousand kilometres. If the final configuration is a stable equilibrium configuration, then the energy released would be:

$$E \approx GM_{\rm core}^2 \left(\frac{1}{R_{\rm final}} - \frac{1}{R_{\rm initial}} \right) \approx \frac{GM_{\rm core}^2}{10\ {\rm km}} \approx 3 \times 10^{53} {\rm erg}. \tag{15.4}$$

This is to be compared with the energy needed to expel the entire envelope of the star. The binding energy of the entire star—core plus the envelope—is just the gravitational potential energy

$$\frac{GM^2}{R} \approx 10^{50} {\rm erg} \tag{15.5}$$

for a $10M_\odot$ star, with a radius of a few million kilometres. Therefore, there is no difficulty in principle to produce a stellar explosion. The energy to expel the envelope is only a very small fraction of the energy released in the collapse of the core.

Let us now try to get a feeling for how the collapse proceeds. The inner part of the core collapses faster than the outer part (see Fig. 15.6). The collapse of the inner core is halted when it reaches the nuclear density of the order of 2.5×10^{14} g cm^{-3}. The infall is arrested due to the resistance of the nuclear matter to further compression. You may recall that while the nuclear force is strongly attractive at nuclear density of the order of 10^{14} g cm^{-3}, it is strongly repulsive at higher density. Actually, the core will *overshoot* the nuclear density and *bounce back*. If the core were perfectly *elastic*

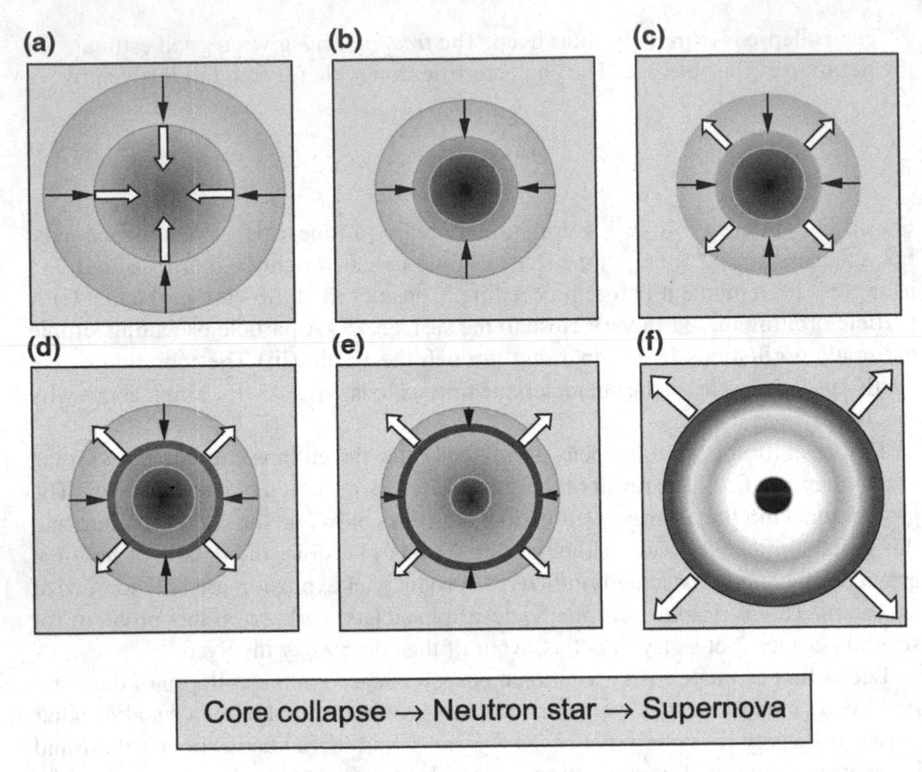

Fig. 15.6 This figure explains how the birth of the neutron star triggers a supernova explosion. The inner part of the core collapses to form a neutron star. The neutron star overshoots its equilibrium radius and bounces back. This leads to the formation of a shock wave. This shock wave encounters the infalling outer core, reverses it motion, and eventually blows apart the rest of the star

it would have enough kinetic energy to bring it back to the original position—but there can be no explosion! To have the huge amount of gravitational potential energy at the star's disposal, the collapsed core must be a stable equilibrium configuration. Otherwise, the kinetic energy of the infall will, once again, be converted to potential energy and we would be back to square one. Imagine you drop a rubber ball from the terrace of your house. As it falls, the potential energy is converted into kinetic energy. If the ball was perfectly elastic it would bounce back into your hand; but it will not have any energy left to impart a vertical momentum to your hand. If the ball was made of clay, for example, it will flatten upon hitting the ground, and you will hear a *thud*. But the ball of clay will stay on the ground. What has happened should be clear. Part of the potential energy has been used up to *deform* the ball and the remaining part heats up the patch of ground, creates sound waves, etc.

In a similar fashion, for the neutron star formed as a result of the infall to be a stable star in equilibrium, energy equal to its *binding energy* (3×10^{53}erg) *would have to be radiated away*. We shall presently argue thatthis energy is released in the form of *neutrinos*. Let us take this for granted for the present, and try to understand the collapse of the core. As we said, the inner part of the core collapses beyond the nuclear density and *bounces back*. As it bounces back, it encounters the infalling material of the outer part of the core. This reverses the infall of the outer core, creating a pressure wave. This pressure wave propagates outwards in an environment of decreasing density. This causes the pressure wave to *steepen* and become a *shock wave*. It had been generally assumed that this shock wave causes the stellar explosion. After all, one needs only a very small fraction of the binding energy of the neutron star to explode the rest of the star.

But actual calculations showed that the shock stalled or fizzled out, instead of expelling the envelope in an explosive manner. In retrospect, the reason is not difficult to understand. You will recall that initially the collapsing core consists mainly of iron. As the inner core collapsed, the nuclei became neutron rich due to inverse beta decay. When nuclear density was reached, the individual nuclei lost their identity and merged to form a nuclear fluid. This is the *neutron star*. The infalling *outer core* (a shell, if you like) is mainly composed of iron. When the outgoing shock wave interacts with the infalling matter, it heats it up to very high temperature. Consequently, the iron nuclei in the infalling shell of matter are *broken up* into free nucleons. The energy for this, of course, comes at the expense of the kinetic energy of the shock wave. As a result, the shock wave has only a small fraction of the initial kinetic energy, and that is not adequate to produce an explosive ejection of the envelope. Obviously, something has to re-energize the shock. According to modern ideas, it is the neutrinos that come to the rescue.

The Trapping of the Neutrinos!

This might surprise you a great deal since neutrinos are supposed to interact extremely weakly with matter. In more technical terms, the *cross-section* for interaction of neutrinos with matter is incredibly small:

$$\sigma_\nu \approx \left(\frac{E_\nu}{m_e c^2}\right)^2 10^{-44}\text{cm}^2 \approx 10^{-44}\text{cm}^2, \tag{15.6}$$

for neutrinos with energy of the order of MeV (Neutrinos produced in nuclear fusion reactions have energy of this order). The above neutrino cross-section is 10^{-18} times smaller than the corresponding cross-section for the interaction of photons with matter. You may perhaps be more familiar with the concept of 'mean free path'. The *mean free path* of the neutrinos in matter is related to the cross-section by

$$l_\nu = \frac{1}{n\sigma_\nu} \tag{15.7}$$

where n is the number of scatterers per unit volume (notice that the mean free path has a the dimension of a 'length'). Multiplying and dividing by the mass of the scatterers (protons and neutrons), the above expression for the mean free path can be expressed in terms of the mass density ρ as

$$l_\nu \approx \frac{2 \times 10^{20}\text{cm}}{\rho} \tag{15.8}$$

where the mass density is in cgs units. For normal stellar matter with density $\rho \approx 1\ \text{g cm}^{-3}$, *the mean free path of the neutrinos is several hundred light years!* This is why it took more than two decades to discover them.

Given this, how can one expect neutrinos to play any role in stellar explosions? What one needs is a piston which is moving out. How can neutrinos possibly act as a *piston*? Well, two things conspire to make the neutrinos a major player.

1. The density of the iron core is roughly 10^{10} g cm^{-3}, and as the core collapses the density becomes even larger. This will drastically decrease the mean free path.
2. The other factor that greatly increases the cross-section for the neutrinos produced during the collapse of the core is their high energy. As we have discussed, the electrons in the collapsing core are highly degenerate and relativistic. The Fermi energy of the electrons is of the order of 10 MeV. The neutrinos produced during the collapse have typical energy of the same order. Notice that the *neutrino cross-section is proportional to the square of the energy of the neutrinos*, measured in units of the rest mass energy of the electrons (refer to Eq. 15.6).
3. Equally important is the huge enhancement of the scattering cross-section that arises due to the *Unified Theory of Weak and Electromagnetic Forces*—the theory due to Salam, Weinberg and Glashow. Let us consider a nucleus with A nucleons. In Fermi's theory of weak interaction, the scattering of the neutrinos by the nucleons inside a given nucleus is *incoherent*. That is, the scattering by the individual nucleons is independent of one another. There is no phase relationship between the scatterers. But in the unified electro-weak theory, the scattering from the nucleus will be *coherent*. In such a situation, *the total scattering cross-section due to a nucleus will be proportional to the square of the number of scatterers,* A^2(and not the number of scatterers). Such a coherent scattering cross-section is given by

$$\sigma_\nu \approx \left(\frac{E_\nu}{m_e c^2}\right)^2 A^2 10^{-45}\ \text{cm}^2. \tag{15.9}$$

For $A = 100$ and $E_\nu \approx 10$ MeV, the coherent scattering cross-section is roughly

$$\sigma_\nu \approx 10^{-39}\text{cm}^2. \tag{15.10}$$

This should be compared with our earlier estimate of 10^{-44} cm^2 (Eq. 15.6)! The coherent scattering cross-section is a hunded thousand times larger.

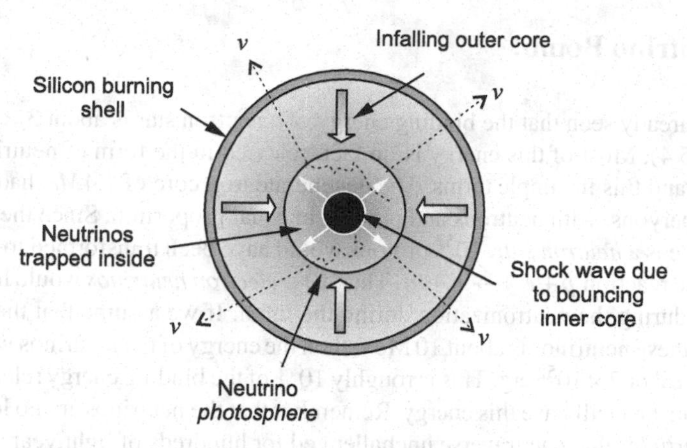

Fig. 15.7 At the density of terrestrial matter, neutrinos have a mean free path of many light years. But the mean free path becomes comparable to the size of the collapsing core when the density reaches 10^{12} g cm^{-3}. Hence, the neutrinos produced during the collapse due to the neutronization of matter, as well as the thermal neutrinos produced by the infant neutron star, will be trapped. They can only diffuse out, till they reach the neutrino photosphere. The trapped neutrinos will exert enormous pressure. Since almost the entire binding energy of the neutron star is released in the form of neutrinos, *neutrinos play a central role in producing the supernova explosion*

This greatly enhanced scattering cross-section, and very high density, implies a mean free path for the neutrinos that is *smaller than the dimension of the core*. Let us estimate the mean free path. We have

$$l_\nu \approx \frac{1}{n\sigma_\nu} = \frac{Am_p}{\rho\sigma_\nu} \approx \frac{2 \times 10^{17}}{\rho}. \tag{15.11}$$

It is instructive to compare Eqs. (15.11) and (15.8), the corresponding expression derived using the *incoherent* scattering cross-section. In the case of independent scattering by the individual nucleons, to convert the number density of scatterers to the mass density we multiplied and divided by the mass of the nucleon. In the case where the nucleons inside a nucleus scatter the neutrinos *coherently*, the scattering entities are the nuclei, rather than individual nucleons. This is why in (15.11) we have multiplied and divided by the *mass of the nucleus* (Am_p) to convert the number density to mass density. The important thing is that mean free path of the neutrinos given by Eq. (15.11) is a thousand times less than estimated with the old physics (Eq. 15.8). Note that at a density of 10^{12} g cm^{-3} the mean free path becomes much less than the size of the inner core. *Thus the collapsing core becomes opaque to neutrinos. The neutrinos are trapped by the collapsing core*! Like the photons inside a star, they can only diffuse out, till they reach an imaginary surface, which we shall call the *neutrino photosphere*, in analogy with the photosphere of the Sun. Once the neutrinos diffusively reach the *photosphere*, they can stream out without any further appreciable scattering (see Fig. 15.7).

The Neutrino Bomb!

We have already seen that the binding energy of a neutron star is about 3×10^{53} erg (see Eq. 15.4). Most of this energy is, in fact, released in the form of neutrinos! Let us understand this in simple terms. The degenerate iron core of $1.4 M_{\odot}$ had roughly 2×10^{57} baryons, with neutrons and protons in equal proportion. Since the result of the collapse is a *neutron star*, 10^{57} protons would have been transformed to neutrons through the reaction $p+e^- \rightarrow n+\nu_e$. Thus 10^{57} *electron neutrinos* would have been produced during the neutronization during the infall. If we assume that the average energy of these neutrinos is about 10 MeV, then the energy of the neutrinos would add up to 10^{64} eV or 2×10^{52} erg. This is roughly 10 % of the binding energy released. But not all neutrinos will have this energy. Remember that the neutrinos are no longer the *elusive particles* that can traverse unchallenged for hundreds of light years. Instead, they are now trapped in the collapsing inner core of the star. Their small mean free path is by virtue of their effective interaction with matter. Recall what happens to photons trapped in a similar fashion in an opaque body. As Kirchoff taught us, they will eventually come to thermal equilibrium with matter. This will now be so for the neutrinos as well. The neutrinos can no longer be regarded as independent particles. They will have to be described by a statistical distribution. This will either be the Boltzmann distribution or the Fermi–Dirac distribution (remember that neutrinos are Fermions) depending upon whether $k_B T \gg E_F$ or $k_B T \ll E_F$. When all this is taken into account, the upshot is that *the electron neutrinos produced during the infall will account for only 1 % of the binding energy of the neutron star.*

What about the remaining 99 % of the binding energy? Remember that our newly formed neutron star will be incredibly hot. Its internal temperature would be about 10^{11} K. At such a high temperature, there will be copious production of neutrinos by the various processes described in Fig. 13.5. *There will be neutrinos and antineutrinos in equal numbers. And neutrinos of all three flavours—electron neutrinos, muon neutrinos and tau neutrinos will be produced with roughly equal probability.* The production of these neutrinos will result in the cooling of the neutron star, just as the emission of black body radiation cools a hot opaque body (cooling due to neutrino emission is the dominant mechanism at the temperatures that obtain in a newly formed neutron star). *Calculations suggest that the cooling time will be very short, ranging from 1 second to a few seconds.* So there will be second burst of neutrino production. The earlier neutrino burst will mostly be electron neutrinos produced during the infall. These neutrinos are produced in a few milliseconds and account for roughly 1 % of the binding energy released. The second burst of neutrino production will be *thermal neutrinos*. These will be produced in a timescale of the order of a second or so, and account for 99 % of the binding energy released.

Remember that all these neutrinos are trapped inside the collapsed core. They will, therefore, exert an enormous outward pressure, just as the stellar plasma and the radiation exert outward pressure in a gaseous star like the Sun. *We thus have a neutrino piston!* It is this pressure due to the trapped neutrinos that energizes the tiring outward moving *shock wave* created by the overshoot and bounce of the newly formed neutron star. To summarize, *the current understanding is that the supernova explosion of a massive star is triggered by a neutrino bomb*!

If you find all this confusing, the following analogy might help to clarify what we have discussed above. Consider a big meteorite hitting the Earth. First there will be tremendous burst of sound waves. Imagine that we want this burst of sound waves—which might even become a shock wave—to do some work, but discover that it does not have enough energy. Not all is lost. Much of the kinetic energy of the meteorite will, in fact, go into heating the patch of earth where it struck. This hot patch will radiate. If this radiation is emitted in a short burst then the major fraction of the kinetic energy of the infalling meteorite will be in this radiation burst. If the conditions are suitable, then such a burst of radiation can do the task you had in mind.

A Guest Star is Born!

Let us summarize, once again, the current understanding of the fate of massive stars. The inner part of the degenerate iron core collapses a little faster and forms a neutron star. The neutron star overshoots its equilibrium radius and bounces back. This generates an outward moving shock wave. What drives the shock is the thermal pressure of the neutrinos which are trapped inside the core. These neutrinos are created during the neutronization of matter in the core, as well as by a variety of processes that dominate at very high temperatures of the order of 10^{11} K. Nearly 99 % of the binding energy of the neutron star are released in the form of these thermal neutrinos.

The outward moving shock blows apart the envelope of the star. Having heated and set in motion the envelope, the shock propagates into the vast space between the stars. When the shock wave exits the star, there will be an ultraviolet flash, or even an x-ray flash. The wavelength range in which the intensity is maximum depends upon the temperature of the shock wave (Recall Wien's Displacement Law). Immediately behind the shock wave will be the heated ejecta. The mass of the ejecta will essentially be the mass of the original star minus the mass of the *stellar remnant* left behind (neutron star of mass $1.4M_{\odot}$). Observations tell us that initially the expansion velocity of the ejecta can be in excess of 10,000 km/s. This is deduced from the Doppler shift of the spectral lines from the hot ejecta. Since the supernova ejecta contains several solar mass of stellar material, such a high velocity implies that the kinetic energy of the expanding ejecta will be roughly 10^{52} erg. This is much more than the Sun will radiate in its entire life!

At the time of the explosion, the ejecta will be opaque (in technical jargon, *optically thick*) and hence will radiate as a black body. As this *fireball* expands, it will remain opaque for a while. And since its surface area will increase with time, the luminosity of the supernova will increase with time (Fig. 15.8). Recall that for a black body

$$\text{Luminosity} = \text{(surface area)} \times \sigma T^4. \tag{15.12}$$

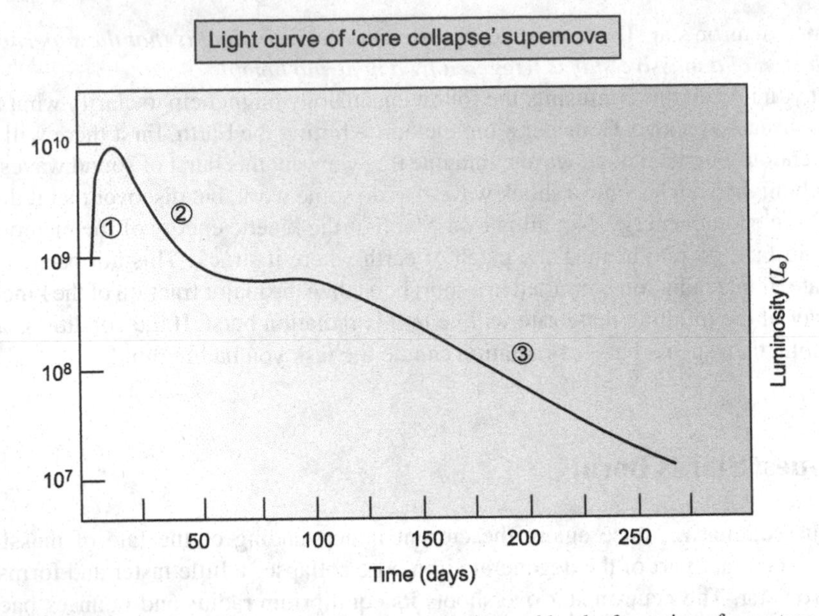

Fig. 15.8 A sketch of the light curve of a supernova triggered by the formation of a neutron star. Initially, the ejecta will be opaque, and hence radiate like a black body. As the ejecta expands, its luminosity will increase because of the increase in the surface area; the temperature of the ejecta does not decrease appreciably. This is the section of the light curve marked (1). The luminosity will start to decline when the ejecta becomes transparent, and also cools (2). If there is no other source of energy, the luminosity of the supernova will decline by many orders of magnitude within about hundred days. But, as may be seen in the figure, the luminosity declines much more slowly. *The energy released in the radioactive decay of cobalt in the ejecta into iron, with a half life of about 77 days, is responsible for the long tail labelled (3)*

After the ejecta expands for a few days (or weeks) it will become transparent (or *optically thin*). From then onwards the luminosity will decrease with time. Once the ejecta becomes transparent, the radiation emitted by it will consist of a *continuum*, with *absorption lines* from the atoms superimposed on it. Since the envelope consists mainly of hydrogen, we would expect to see strong absorption features corresponding to the *Balmer lines*. These lines are, indeed, seen very prominently in the spectrum of light from the supernova. After about hundred days the supernova light curve declines *exponentially*; this is the long tail in Fig. 15.8. During this phase, the main source of energy which powers the light curve is the radioactive decay of the unstable isotope of Cobalt. Earlier, when we discussed the termination of the exothermic nuclear fusion reactions, we said that the final product will be $^{56}_{26}$Fe since it has the largest binding energy. Strictly speaking, we should have said that the final *stable* nucleus will be $^{56}_{26}$Fe. Some radioactive nickel ($^{56}_{28}$Ni) will also be produced. This isotope of nickel has 28 protons. This nickel will decay to cobalt, which, in turn, will decay to $^{56}_{26}$Fe. The decay will proceed as follows:

$$^{56}_{28}\text{Ni} \rightarrow {}^{56}_{27}\text{Co} + e^+, \quad (\text{half life} \ = \ 6 \text{ days})$$
$$^{56}_{27}\text{Co} \rightarrow {}^{56}_{26}\text{Fe} + e^+, \quad (\text{half life} \ = \ 77 \text{ days}) \qquad (15.13)$$

The energy released in the radioactive decay of cobalt to iron is what powers the light curve at later times. You may ask why is there nickel and cobalt in the supernova ejecta? Firstly, the core will still contain the unstable isotopes of nickel at the time of implosion. This is because the final phase of silicon burning, during which the iron group of elements are synthesized, lasts only a day or so. Second, recall that the outermost layers of the degenerate *iron core* were expelled by the outward moving shock. Therefore, the ejecta will contain some iron and nickel.

Historically, only visible radiation was detected from supernovae. Today, one has been able to detect prompt radio and x-ray emission from supernovae. Since the emission mechanism is different at different wavelengths, a multiwavelength study of supernovae will shed much light not only on the nature of the explosion, but also on the abundance of elements synthesized, the acceleration of cosmic rays during the explosion, etc. This is an extremely active field today.

Although the light from the supernova will fade away in a few months, the blast wave, and the ejecta, will continue to expand in the vast space between the stars. Initially, the ejecta will expand freely according Newton's law of motion. As the blast wave sweeps up more and more matter in the interstellar space, the ejecta will slow down, just as a *bulldozer* slows down as it ploughs more and more earth. Finally, after several thousand years, the expanding blast wave will come to rest. This will happen when the thermal pressure inside the cavity excavated by the blast wave becomes equal to the ambient pressure of the interstellar medium. By that time, the initial kinetic energy of the ejecta would have been deposited into the interstellar medium. Remember that it is great deal of energy, of the order of 10^{52} erg. Since a massive star explodes in our Galaxy every 30 or 40 years, the expanding blast waves cause havoc in the interstellar medium. They heat up the gas to millions of degrees; they accelerate interstellar clouds of gas; they trigger the collapse of giant clouds, leading to the birth of new stars. Thus, *there is a symbiotic relation between the birth and death of stars!*

Diamonds are not for Ever!

The explosions of massive stars that we have been discussing are classified as **Type II supernovae**. Their main characteristic is that the ejecta is rich in hydrogen, and they leave behind a neutron star. There is another type of supernovae which do not show any hydrogen in its spectrum, and which do not leave behind any stellar remnant. These are known as **Type Ia supernovae**.

Till the 1980s, it was believed that these supernovae are the explosions of intermediate mass stars when they ignite carbon in their degenerate cores. Since the whole star is blown apart, no stellar remnant will be left behind. But this scenario had to be abandoned when white dwarfs were discovered in young open clusters of stars. We discussed this in detail in the preceding chapter. You will recall that according to present understanding, all stars with mass up to about $9M_{\odot}$ will end their lives as white dwarfs. Therefore, the single star scenario for Type I supernovae has now been abandoned.

Accreting White Dwarfs in Close Binary Systems

According to the currently prevalent opinion, *Type I supernovae are the result of white dwarfs being pushed over the Chandrasekhar limiting mass.* One of the scenarios in which this can happen is the following. The majority of stars we see in the sky are, in fact, binary stars—two stars going around a common centre of mass. In some cases, the more massive of the two stars could have ended its life as a *white dwarf.* So we will have *a binary consisting of a white dwarf and a gaseous star.* At some stage, the companion star will also evolve and become a *giant.* If the binary is very tight, tidal forces could result in matter from the outer layers of the giant star being torn apart from the parent star and accreting onto the white dwarf. If this mass accretion onto the white dwarf lasts long enough, then the mass of the white dwarf could eventually grow to the Chandrasekhar limit, resulting in a collapse. It is not at all clear whether the collapse will just lead to the formation of a neutron star or result in the detonation and destruction of the white dwarf. Remember that the mass limit for *both* these possibilities is roughly $1.4M_{\odot}$. As we discussed in the previous chapter, the effectiveness of the cooling by the neutrinos can tip the balance either way.

But there is a fundamental difficulty with this scenario. Most white dwarfs have a mass of around $0.6M_{\odot}$. This implies that a white dwarf in a binary would have to accrete about $0.8M_{\odot}$ from its companion before it reaches the Chandrasekhar limiting mass. Observations of accreting white dwarfs–from which we can infer the typical accretion rate–tell us that it would take about a billion years or more to accrete such a large amount of mass. So there is a dilemma.

- *The companion star must be massive enough to donate a large amount of mass to the white dwarf.*
- *And yet, it must live long enough to actually donate nearly a solar mass worth of material to the white dwarf!*

Remember that the more massive a star is, shorter is its life. This is the dilemma.

Even if assume that the white dwarf has a suitable and obliging companion, it is not at all obvious that it will be able to *accept and retain* this mass. The difficulty is the following. It is clear that the accreting matter, ripped apart from the outer layers of the giant star, would be mostly hydrogen. *When this hydrogen settles on the surface of the white dwarf, it will be compacted to a very high density, and very high temperature, due to the strong surface gravity of the white dwarf. Thus the accreted gas will find the conditions just right for hydrogen to fuse to helium.* Recall that this is precisely what happens in the hydrogen burning shell surrounding the inert

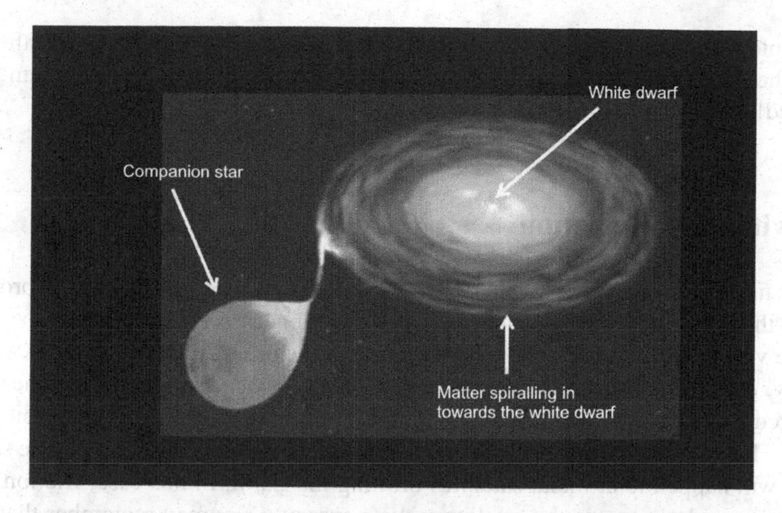

Fig. 15.9 Accreting white dwarf in a binary system. Matter from the outer layers of the companion is ripped out due to the strong gravity of the white dwarfs. This matter cannot directly fall onto the white dwarf. This is because it has angular momentum, by virtue of the fact that the two stars are going around the common centre of mass. The matter can only fall in by spiralling in. In the process, an *accretion disk* is formed around the white dwarf. When matter finally reaches the surface of the white dwarf, there is no guarantee that it will stay there! If the accreted matter burns explosively, then it will be ejected from the surface. Such mass ejections are identified with the Nova phenomenon

helium core of a red giant star. But there is an important difference. In the present case, there is nothing equivalent to the *envelope of the star* sitting on top of the accreted matter. Consequently, the energy released in the process (the synthesis of the accreted hydrogen into helium) will *blow off* the accreted matter. This is a quite common occurrence and is identified with the phenomenon of *Novae*—as opposed to *Supernovae*. It is, therefore, very unlikely that the accreting white dwarf will grow in mass and reach the Chandrasekhar limit. The only possibility for this to happen is if matter accretes at an unrealistically large rate. In that case, the energy released by the fusion reaction on the surface will not be able to *lift* and *expel* the accreted matter. Detailed calculations show that while this is possible in principle, it is very unlikely.

This raises the following interesting question: *is there any other scenario for a white dwarf to grow in mass, and get pushed over the Chandrasekhar limit?*

Coalescence of White Dwarfs

Yes! Occasionally, the evolution of binary stars will lead to a *white dwarf binary*; both stars ending their lives peacefully as white dwarfs (Fig. 15.9). Let us assume that they are both carbon–oxygen white dwarfs; after all, they are the most common variety. Now we will leave it to Einstein's theory of gravity to make the two white dwarfs spiral in and eventually merge. If the total mass of the two white dwarfs equals or exceeds the Chandrasekhar limit, then carbon will ignite and we will have an explosion. You will ask 'Why should they spiral in? After all, the Earth and the other planets have been safely orbiting the Sun ever since the solar system was formed!' Let us try to understand the underlying physics.

Gravitational Radiation

Let us first discuss the classical physics problem of an electron going around a proton. Since the electron is in a curved orbit, it will experience *acceleration* equal to v^2/r, where v is the orbital speed and r is the radius of the orbit. According to Maxwell's theory, *an accelerating electron will radiate electromagnetic radiation.* Since the radiated energy can only come at the expense of the orbital energy, the orbit will shrink. In other words, the electron will spiral into the proton (By the way, the same thing will happen to artificial satellites orbiting the Earth. In this case, friction due to the atmosphere causes the satellite to lose energy). You may remember that J.J. Thompson's model of the hydrogen atom had to be abandoned because of this—the atom would not be stable. Niels Bohr solved that problem by introducing quantum mechanics.

In a similar fashion, a body of mass m_1 going around another body of mass m_2 will emit *gravitational radiation.* In Newton's theory of gravity, there is no such thing as gravitational radiation. But Einstein's theory—the General Theory of Relativity—predicts the existence of gravitational radiation. Like electromagnetic waves, gravitational waves will also propagate at the speed of light. Since the energy radiated in the form of gravitational waves will have to come at the expense of the orbital energy (like in the example mentioned in the preceding paragraph), the orbit will shrink, and the two bodies will eventually coalesce. The question is one of timescale. The energy radiated as gravitational waves per unit time depends upon the masses and the radius of the orbit a in the following manner:

$$L = -\frac{dE}{dt} = \frac{32}{5}\left(\frac{G^4}{c^5}\right) m_1^2 m_2^2 (m_1 + m_2) \frac{1}{a^5}. \tag{15.14}$$

The larger the mass, the larger is the luminosity. The smaller the orbit, the larger is the luminosity. For a circular orbit, the rate of shrinking of the orbit is given by the formula:

$$\frac{da}{dt} = -\frac{64}{5}\left(G^3/c^5\right) m_1 m_2 (m_1 + m_2) \frac{1}{a^3}. \tag{15.15}$$

As the orbital separation decreases, the rate of decrease increases. So the two stars will spiral into each other at an ever-increasing rate (see Fig. 15.10).

There has been no direct detection of gravitational waves, as yet. Several very sophisticated experiments are under way to detect these waves, but success is still some years away. In the mean time, there is indirect, but very compelling, evidence for the existence of these waves. There are a couple of *double neutron star systems*

Fig. 15.10 An artist's impression of two white dwarfs spiralling towards each other. Such a close binary of two white dwarfs has actually been found by the Chandra X-ray Observatory. The system known as J0806 is roughly 1600 light years away. Incredibly, *the orbital period of the binary is only 321seconds*! Such a close binary will rapidly spiral in due to the emission of gravitational radiation, and eventually coalesce. The result will be a Supernova of Type I [Credit: NASA/Tod Strohmayer (GSFC)/Dana Berry (Chandra X-ray Observatory)]

with extremely small orbital separation. The orbital period is only a couple of hours, and the orbital separation is less than a solar radius! The important point is this. *The orbit of these double neutron star systems is observed to be shrinking. And the rate at which the semimajor axis is shrinking agrees spectacularly well with the prediction made by Einstein's theory* (see Eq. 15.15). The scientific community is so convinced of this that J. H. Taylor and R.A. Hulse, the discoverers of the first double neutron star system in which this effect was first seen, were awarded the Nobel Prize for physics. Faced with this compelling evidence, we must conclude that gravitational waves do exist. It is only a matter of time before they are actually detected in a direct experiment.

Let us return to our story of the double white dwarf binary. If the initial orbit of such a system is sufficiently tight, then one can hope for a merger in a timescale $\leq 10^{10}$ years. The merger will not only increase the mass to the Chandrasekhar limit, it will also result in a considerable heating. Both these will help the carbon detonation scenario, resulting in a Type I supernova. This is the current scenario for these rare supernovae.

An interesting thing about Type I supernovae is that their luminosity at maximum is the same. In other words, *they are standard candles*. It is this fact that makes them so invaluable in cosmological studies. Indeed, *it is the detection of these standard candles in distant galaxies that has enabled astronomers to conclude that the Universe is not only expanding, but the expansion is accelerating*.

Black Holes

The main theme of this chapter has been the collapse of the degenerate iron core that forms at the end of the nuclear cycles. We have argued that the result of the collapse will be a neutron star of mass very nearly equal to 1.4 $M_\odot$. The neutron star overshoots its equilibrium radius and bounces back. The shock wave due to this, aided by the burst of neutrinos from the cooling of the infant neutron star, produces the supernova explosion.

In some cases, a stable neutron star may not be the end product of the collapse, and the result of the collapse may be a black hole. This can happen due to a variety of reasons.

1. The shock may stall, despite help from the neutrinos. In such cases, there could be a substantial infall of matter onto the newly formed neutron star. If this accreted matter gets neutronized, and increases the mass of the neutron star to the limiting mass for neutron stars (roughly two solar mass) then there will be a further implosion, resulting in a black hole.
2. If the neutron star matter is not stiff enough, then there will be no bounce back at all. This can happen if the particles such as π mesons and K mesons spontaneously occur near the core of the neutron star. Unlike the neutron and the proton, these mesons obey Bose–Einstein statistics. Since they do not have to obey Pauli's exclusion principle, they would all condense into the zero momentum state. Consequently, their contribution to the pressure would be *zero*. This would render the neutron star matter rather *soft*. Under such circumstances, the neutron star would be unstable and continue to collapse. The collapsing core will directly form a black hole.

The details of how black holes form from the collapse of massive stars are not clear at the moment. But there is mounting evidence for the existence of stellar mass black holes. This leads us to conclude that at least in some cases, the end product of the evolution of massive stars will be black holes. And on that note, let us conclude our review of the life history of stars.

Epilogue

We began with a historical perspective of the end-states of stars. Our story began with the discovery of the companion of Sirius, and the startling realization that its mean density must be roughly a million times the density of our Sun. Eddington feared that such stars will find themselves in an awkward predicament when their supply of subatomic energy is exhausted. He argued that such stars will not be able to save themselves by expanding. This paradox was resolved in 1926 by R. H. Fowler by invoking the newly discovered Fermi–Dirac statistics. It is remarkable that this was the very first application of the new quantum statistics. It is quite extraordinary that the stability of a body as large as a star is to be understood in terms of electrons having to obey Pauli's exclusion principle! This prescient suggestion by Fowler was followed up young Chandrasekhar, who constructed a complete theory of white dwarfs. He concluded that all stars, regardless of their mass, will ultimately find peace as white dwarfs.

Unfortunately, this sense of security did not last long. Chandrasekhar himself found the flaw in the above conclusion. During his long voyage to England in 1930, he made the startling discovery that stable white dwarf configurations are not possible for stars with masses higher than $1.4 M_{\odot}$. By 1934, he had established that $1.4 M_{\odot}$ is the limiting mass for white dwarfs. Stars more massive than this cannot be supported against gravity by the degeneracy pressure of the electrons.

This raised the following basic question. What is the fate of stars more massive than the Chandrasekhar mass limit for white dwarfs? Chandrasekhar had already found an answer to this in 1932. He showed that if the mass of a star exceeded a certain critical mass, matter will never become degenerate however large the density may become. Clearly, such stars cannot be saved from gravitational collapse by appealing to Fermi–Dirac statistics. Based on this conclusion, Chandrasekhar made the bold prediction that sufficiently massive stars will collapse to form *singularities*.

The fate of stars with mass greater than $1.4 M_{\odot}$, but less than the above-mentioned critical mass, was still unclear. The answer to this question came with the discovery of the neutron in 1932. In 1937, Landau argued that the collapse of a star will eventually be halted when the electrons combine with the protons to form neutrons. When the

G. Srinivasan, *Life and Death of the Stars*, Undergraduate Lecture Notes in Physics,
DOI: 10.1007/978-3-642-45384-7, © Springer-Verlag Berlin Heidelberg 2014

density of stellar matter reaches the nuclear density $\sim 10^{14}$ g/cm^3, the degeneracy pressure of the neutrons will arrest the gravitational collapse.

The concept of a *neutron star* stimulated Oppenheimer and Volkoff to investigate whether there was a maximum mass for neutron stars, just as there is maximum mass for white dwarfs. In 1938, they came to the conclusion that it is not possible to have stable neutron stars if their masses exceeded roughly $0.7 M_\odot$. But this mass limit for neutron stars was not to be regarded as an exact result. This was due to an inadequate understanding of nuclear forces at the time. The important conclusion was that there will be maximum mass for the neutron star.

This discovery led Oppenheimer and his student Snyder to study the collapse of a massive star within the premise of Einstein's General Theory of Relativity. In a remarkable paper published in 1939, they came to the conclusion that sufficiently massive stars will collapse to form *black holes*. Although they did not highlight it, in the General Theory of Relativity, a star that collapses to form a black hole has no option but to continue to collapse till, eventually, it becomes a *spacetime singularity*.

These theoretical predictions regarding the ultimate fate of stars were made by 1939. There was one additional prediction concerning stars that collapse to form neutron stars. And this extremely prescient prediction was made by Baade and Zwicky in 1934. They advanced the revolutionary idea that when a star collapses to form a neutron star at its centre, the gravitational binding energy released will produce a spectacular stellar explosion. They hypothesized that this may be the origin of the supernovae.

It would be instructive to gather together the explicit and implicit predictions that were made by 1939, based upon very general considerations:

1. The maximum mass of stars that will find peace as white dwarfs will be $1.4 M_\odot$.
2. Stars more massive than this would continue to collapse, and eventually find equilibrium as neutron stars.
3. Their birth will be accompanied by a burst of neutrinos and a supernova explosion.
4. The masses of neutron stars will be very nearly equal to the Chandrasekhar mass limit for white dwarfs, namely, $1.4 M_\odot$.
5. In stars above a certain critical mass, radiation pressure will prevent degeneracy from setting in. Consequently, such stars will become black holes, and a collapse to a singularity is inevitable.

An examination of Fig. E.1 will convince you that the modern conclusions concerning the end states of stars are remarkably similar to the above mentioned conclusions! You will notice that the upper value of the initial mass of the star that will end up as a white dwarf is larger than the Chandrasekhar limiting mass for white dwarfs. This is because stars lose a substantial amount of mass before ending their lives. But this was not known in the 1930.

As for the critical mass above which a star will become a black hole, it is still uncertain!

Modern observations have also spectacularly confirmed the three other predictions made in the 1930s.

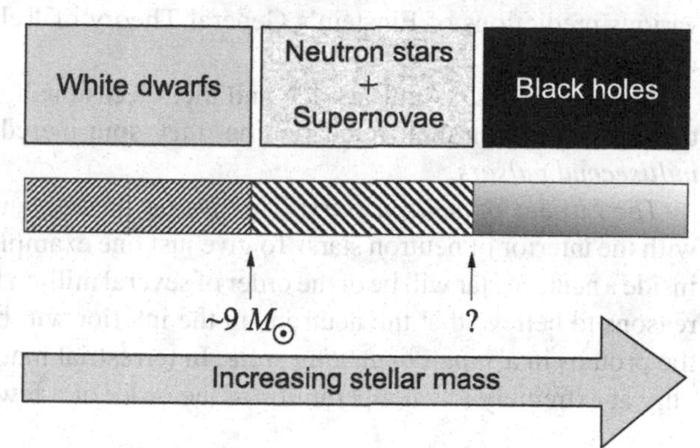

Fig. E.1 A summary of our current understanding of the ultimate fate of stars

1. The presence of a neutron star at the centre of the Crab Nebula, the remnant of the supernova explosion of 1054 AD, as well as in many other supernova remnants, firmly establishes the connection between neutron stars and supernovae.
2. The detection of the burst of neutrinos on 23 February 1987 from the supernova in the Large Magellanic Cloud finally provided the definitive proof for the conjecture by Baade and Zwicky that the birth of a neutron star triggers the supernova, as well as Landau's conjecture of neutronization of matter at very high densities.
3. The fact that the measured masses of neutron stars is invariably very close to $1.4 M_\odot$ is a spectacular confirmation of the existence of the Chandrasekhar Limit, a result which Eddington had rejected! And on that note we shall end this monograph.

A Sneak Preview

The next volume in this series, entitled ***Neutron Stars and Black Holes***, will be devoted to the physics and astrophysics of neutron stars and black holes. Here is a partial list of the topics that will be discussed.

Pulsars: Neutron stars manifest themselves in two different ways. Solitary neutron stars, rapidly rotating, and endowed with incredibly large magnetic fields of the order of thousand million gauss, are detected as pulsating stars. They are known as pulsars. This population of neutron stars is mainly detected through the radio waves they emit. Interestingly, this radio emission is coherent, just the light emitted by a laser is coherent.

Binary Neutron Stars: Although smaller in number compared to radio pulsars, the population of neutron stars in binary systems are wonderful laboratories for a variety of physical processes. Neutron stars accreting matter from gaseous companion stars manifest themselves as extremely luminous x-ray sources. Double neutron star systems with extremely small orbital separation are wonderful laboratories for verifying various predictions of Einstein's General Theory of Relativity with unprecedented accuracy.

Recycled Pulsars: Pulsars age and die. Occasionally, they are resurrected from their graveyard. In their reincarnation, they spin incredibly rapidly. These are the *millisecond pulsars*.

The Physics of neutron stars: There is fascinating and exotic physics associated with the interior of neutron stars. To give just one example, although the temperature inside a neutron star will be of the order of several million kelvin, there are compelling reasons to believe that the neutrons in the interior will be in a *superfluid state*, and the protons in a *superconducting state*. In terrestrial matter, these phenomena occur only at extremely low temperatures of the order of a few kelvin!

Black Holes

There is now compelling observational evidence for the existence of black holes. In the current volume we discussed black holes which are the end products of stellar evolution. We shall discuss this evidence. One of the paradigms of contemporary astronomy is that almost every galaxy has a giant black hole at its centre. These supermassive black holes, ranging in mass from a million solar mass to a billion solar mass, are the central engines that power the quasars. We shall discuss the formation of these black holes, as well as the physical phenomena that occur in their vicinity.

Although the General Theory of Relativity attracted the attention of the physicists soon after it was published, it was mainly in the context of cosmology. Very few astronomers believed that the theory would be relevant for astronomical bodies. Remember, not many astronomers or physicists took serious note of the great discoveries by Chandrasekhar and Oppenheimer. But things changed with the discovery of neutron stars and quasars. There was a renaissance. The second half of the twentieth century was the golden age of General Relativity. During this period, many important theorems were proved concerning the properties of black holes. These developments culminated in Hawking's discovery that particles can come out of black holes. *Black holes can evaporate!* In this discovery, one saw glimpses of a fusion between Einstein's theory of gravity and quantum mechanics. A qualitative description of these exciting developments will form an important part of the next volume.

Suggested Reading

1. G Venkataraman, *Chandrasekhar and His Limit*, Universities Press (India), 1992.
2. K C Wali, *Chandra*, Penguin Books India, New Delhi, 1990.
   Originally published by the University of Chicago Press, Chicago, 1990. This is a magnificent biography of Subrahmanyan Chandrasekhar.
3. S Chandrasekhar, *Truth and Beauty*, University of Chicago Press, Chicago, 1987. This is a collection of some of Chandrasekhar's famous public lectures, including his classic lecture entitled *Shakespeare, Newton and Beethoven or Patterns of Creativity*. I strongly recommend this volume.
4. G Venkataraman, *At the Speed of Light*, Universities Press (India), 1993.
5. *Frontiers in Astronomy – Readings from Scientific American*. W H Freeman, San Francisco, 1970.
   This has a wonderful collection of articles from Scientific American.
6. L Murdin, and P Murdin, *Supernovae*, Cambridge University Press, Cambridge, 1985.
7. G Venkataraman, *Bose and His Statistics*, Universities Press (India), 1992.
8. G Venkataraman, *Quantum Revolution 1: The Breakthrough*, Universities Press (India), 1993.
9. G Venkataraman, *Quantum Revolution 2: QED—The Jewel of Physics*, Universities Press (India), 1993.
10. G Venkataraman, *Quantum Revolution 3: What is Reality?*, Universities Press (India), 1993.

G. Srinivasan, *Life and Death of the Stars*, Undergraduate Lecture Notes in Physics,
DOI: 10.1007/978-3-642-45384-7, © Springer-Verlag Berlin Heidelberg 2014

Index

G. Srinivasan, *Life and Death of the Stars*, Undergraduate Lecture Notes in Physics,
DOI: 10.1007/978-3-642-45384-7,